Bodily Integrity and the Politics of Circumcision

Bodily Integrity and the Politics of Circumcision

Culture, Controversy, and Change

Edited by

George C. Denniston

University of Washington,
Seattle, Washington, U.S.A.

Pia Grassivaro Gallo

Faculty of Psychology,
University of Padua, Italy

Frederick M. Hodges

Berkeley, California, U.S.A.

Marilyn Fayre Milos

National Organization of Circumcision
Information Resource Centers,
San Anselmo, California, U.S.A.

and

Franco Viviani

Faculty of Medicine and Faculty
of Psychology,
University of Padua, Italy

Springer

A C.I.P. Catalogue record for this book is available from the Library of Congress.

ISBN-10 1-4020-4915-3 (HB)
ISBN-13 978-1-4020-4915-6 (HB)
ISBN-10 1-4020-4916-1 (e-book)
ISBN-13 978-1-4020-4916-3 (e-book)

(BS/DH)

9 8 7 6 5 4 3 2 1

springer.com

Dedication

For Jeannine Parvati Baker, in loving memory.

CONTENTS

PREFACE

How is it possible that at the beginning of the twenty-first century it is still controversial to suggest that males and females should be allowed to keep all of the body parts with which they are born? Medical science is more advanced today than it has ever been in the history of mankind, and yet the medical profession still tolerates within its ranks the presence of doctors who use the pretense and mask of "science" to advocate and perform an ancient tribal blood rite that amputates and destroys the healthy genital tissue from innocent individuals who are powerless to escape. These individuals are then compelled to spend the rest of their lives with truncated, scarred, blunted, and impaired genitals — a fate they did not choose but one that was forced upon them by a perpetrator driven by something other than the best interest of the child.

Clearly, part of the problem is that the ethical standards that are supposed to regulate the practice of medicine were designed by the medical industry itself or by lay people under its sway. For instance, no US hospital ethics committee has ever consulted with genital integrity proponents when setting policy over the practice of circumcision. Instead, these committees rely entirely on the persuasions of anti-foreskin militants or lay people who have been conditioned to accept circumcision as "harmless" or as "the parents' choice." US hospital ethics committees even refuse to follow world ethical guidelines for circumcision.[1] As a perfectly apt analogy, imagine if mobsters were allowed to frame the ethics and draft the legislation regulating racketeering: Mobsters would paint themselves as ethical benefactors and present racketeering as a benefit to society and to its victims. So it is with circumcision: the circumcisers have thus far been given free reign to set policy and practice standards regarding circumcision. Obviously, they have a conflict of interest.

Another significant part of the problem is that the alleged medical "facts" about the results of circumcision are under the control of circumcisers. Every medical "study" promoting circumcision as beneficial has been crafted by a representative of the circumcision industry. Even the official position statements on circumcision issued by US medical organizations, such as the American Academy of Pediatrics, have wholly or in part been drafted by circumcision proponents.[2] As a perfect analogy, imagine what the official position of a hypothetical "American Academy of Pediatrics Task Force on

Smoking" would be if representatives of the tobacco industry and scientists in their pay were invited to sit on the committee or even serve as its chairman. The public would doubtless be told that smoking has potential medical benefits and that, besides, it is an ancient tradition among many indigenous peoples of the world, and therefore demands respect and tolerance. So it is with circumcision.

Furthermore, a still more insidious part of the problem is that the victims of circumcision are generally conditioned to be ignorant of the harm inflicted upon them. Many victims of infant male circumcision in the United States have never seen an intact penis. They do not understand how an intact penis functions. They will never be able to comprehend or imagine the wider spectrum of sensations that have been permanently lost because of the destruction of the foreskin. Consequently, the victims believe the misinformation and disinformation promoted by the circumcision industry, insinuating that the intact penis has no more functions or sensations than those performed by the circumcised penis. The victims often believe that the scars on their penis are natural anatomical features common to all penises. The desiccated, disfigured, and discolored glans that has been artificially externalized as a result of circumcision seems normal and acceptable to the vast majority of circumcised American males because the circumcision industry has convinced them that these pathological conditions are "normal" and even preferable. When circumcised Americans do have the rare opportunity to see undamaged genitals, they sometimes recoil in revulsion — exactly as they have been programmed to do. The tragic universal human tendency to prefer the familiar over the superior has served the interests of circumcisers quite well.

The scope of the problem is vast. Nearly all Muslim and Jewish boys around the globe are circumcised for "religious" reasons — as if faith and religious conviction could be carved into a baby with a knife. This illogical notion remains unchallenged because the true reasons that these parents allow their sons to be circumcised are based on fears of social ostracism and expulsion from the community. Boys and girls are also circumcised in various sub-Saharan African countries. The excuses offered to justify these acts are strikingly similar to those historically and currently offered by the medical profession in the United States: the maintenance of moral purity, cleanliness, conformity, and protection from a long list of incurable and fatal diseases.

Currently, 1.1 million American baby boys each year are subjected to circumcision. The United States is the only Western nation that tolerates routine neonatal circumcision. While no European countries have ever had a tradition of male circumcision, during the twentieth century the United States came under the influence of forces and circumstances that enabled a small band of circumcision proponents to influence the medical profession and launch their attack on the normal American penis. While the alleged medical reasons invented to justify circumcision constantly change in synchronicity with the shifting anxieties of the American public, the underlying motives of the circumcisers remain unchanged. Indeed, seen in this light, circumcision is a willful destruction of the intact penis and a disregard for the cultures that respect and protect genital integrity.

Leaders within the growing worldwide movement to abolish circumcision have for many years grappled with the fact that the language and vocabulary traditionally used to describe and denote the surgical destruction of the genitals has been under the control of circumcisers, whose goal is obfuscation rather than accuracy. It was indeed a great moment of sudden insight and clear-sightedness when human rights activists realized that the intentionally feeble and inaccurate word "circumcision" should be replaced with the precise term "genital mutilation." Amputation of the penile foreskin and the clitoral hood are, in fact, mutilations. When the mind is freed from the influence of circumcisionist propaganda and can make a sober comparison between circumcised and intact genitals, it is immediately clear that circumcised genitals have been damaged.

Still, few people want to be told that they are mutilated. The problem for human rights activists has been to find a balance between clinical accuracy and social acceptability in terms of the vocabulary they use. Many compromises have been made. Probably, no one term will ever serve all purposes and, depending upon the situation, the terms probably should be used interchangeably, always with respect for the survivors of genital mutilation. In the 1980s, when the issue of female circumcision in Africa became a lightening-rod topic among political activist circles in the United States, the strident approach adopted by American feminists merely served to offend and alienate African women rather than shame African men, who were mistakenly viewed as the perpetrators of female circumcision. It was African women who rejected the offensive terms and unwelcome politics espoused by American feminists and substituted better vocabulary based on an understanding of African cultural realities. Many of the papers in this book reflect the experiences and insights of Africans themselves, and they trace the transformation of the worldwide movement to promote genital integrity and abolish circumcision, thanks to the efforts of African activists together with European activists who recognized the wisdom of what African women were telling them.

The problem of male and female circumcision, wherever it occurs in the world, is one of the most difficult and complex problems ever to face humanity. While slavery has not been entirely abolished throughout the world, it is at least recognized as a crime, and those who perpetrate it properly understand that they are involved in a criminal activity. If humanity as a whole could agree that slavery had to be abolished despite the powerful economic advantages it offers to slave holders, circumcision can surely be abolished. It took about a hundred years to abolish slavery. It may take as long to eradicate circumcision and create a world where all humans are permitted the dignity of keeping all of the body parts with which they were born. If this calculation is correct, then this bodes well for the future. Legal and even medical opposition to male and female circumcision is rapidly increasing. The movement to rid the world of genital mutilation is about thirty years old. If the momentum of enlightenment and ethical reform continues at its same pace, children born today should witness the eradication of circumcision within their lifetimes. A new era of freedom and dignity will have been born.

George C. Denniston

Frederick M. Hodges

Marilyn Fayre Milos

REFERENCES

---

1. Geisheker JV, Hill G. US bioethics fall short of world standards. *Arch Dis Child Fetal Neonatal Ed* 2004;89:563.
2. Goldman R. Circumcision policy: a psychosocial perspective. *Paediatr Child Health* 2004;9(9):630-3.

ACKNOWLEDGMENTS

To the people who helped make the Eighth International Symposium on Circumcision and Human Rights, held at the University of Padua in Padua, Italy, a success, we extend our heartfelt thanks. Special thanks are owed to Dr. Pia Grassivaro Gallo, Dr. Franco Viviani, The Padua Working Group on Female Genital Mutilation, Ken Brierley, Sheila Curran, and Dr. Peter Ball. In addition, we would like to express our appreciation to the contributors for allowing us to publish their work. We would also like to express our deepest gratitude to Dr. James L. Snyder for his immeasurable encouragement and generous support.

CONTRIBUTORS

Peter J. Ball, MA, MB, Bchir, Vice Chairman, National Organization of Restoring Men – UK. Turnbridge Wells, Kent, England.

Jim Bigelow, PhD, (Claremont Graduate School), author of *The Joy of Uncircumcising*. First published in 1992, it is now in its third edition. Pacific Grove, California, USA.

Dan Bollinger, Director, NOCIRC of Indiana and Director, International Coalition for Genital Integrity. West Lafayette, Indiana, USA.

Lisa Capparotto, Padua Working Group on Female Genital Mutilation, Department of General Psychology, University of Padua, Italy.

Silvia Casale, University of Florence Research Centre for Preventing and Curing Female Genital Mutilation and its Complications. Florence, Italy.

Lucrezia Catania, Department of Gynecology, University of Florence, Italy.

Gian Luca Costardi, Padua Working Group on Female Genital Mutilation, Department of General Psychology, University of Padua, Italy.

Amber Craig, MA, Director, NOCIRC of North Carolina and coordinator of the Medicaid Project to bring an end to public funding of non medically indicated surgery of non-consenting minors. Chapel Hill, North Carolina, USA.

Nagla Dawelbait, holds a degree in Agriculture from the University of Karthoum and is currently enrolled in a graduate course in Forest Ecology, Faculty of Agriculture, University of Padua, Italy.

George C. Denniston, MD (Princeton University), MPH (Harvard School of Public Health), founder and Director of Doctors Opposing Circumcision (D.O.C.), co-author, *Doctors Re-Examine Circumcision*, co-editor, *Sexual Mutilations: A Human Tragedy, Male and Female Circumcision: Medical, Legal, and Ethical Considerations in Pediatric Practice, Understanding Circumcision: A Multi-Disciplinary Approach to a Multi-Dimensional Problem,* and *Flesh and Blood: Perspectives on the Problem of Circumcision in Contemporary Society,* and Clinical Assistant Professor, Department of Family Medicine, University of Washington. Seattle, Washington, USA.

Lara Franco, Padua Working Group on Female Genital Mutilation, Department of General Psychology, University of Padua and University of Naples, Italy.

John V. Geisheker, JD, LL.M, Attorney at Law, General Counsel, Doctors Opposing Circumcision. Seattle, Washington, USA.

Leonard B. Glick, MD (University of Maryland), PhD (University of Pennsylvania), Professor of Anthropology, Hampshire College, Amherst, MA (retired), author of *Abraham's Heirs: Jews and Christians in Medieval Europe* (1999), "Religion and Genocide," in I.W. Charney, ed., *The Widening Circle of Genocide* (1994), *Marked in Your Flesh: Circumcision from Ancient Judea to Modern* America (2005), and articles on religion and ethnicity. His book, *Circumcision, Jewish Ethnicity, and the Jewish-Christian Encounter*, is in preparation. New Salem, Massachusetts, USA.

Nansi Glick, MA, received her degree in Anthropology and is an academic bookseller. New Salem, Massachusetts, USA.

Pia Grassavaro Gallo, PhD, Associate Professor in general biology, Faculty of Psychology, University of Padua. Her research work for the last twenty-seven years has been dedicated to Somali women and to the physical development of children in Somalia. At the invitation of the Somali Ministry of Public Health (1981), she was invited to take part in a scientific mission to Somaliland. Her scientific activity in Somalia has received recognition both at national and international levels. Padua, Italy.

R. Wayne Griffiths, MS, MEd, sociologist, educator, and co-founder of NORM (National Organization of Restoring Men). Concord, California, USA.

Frederick Mansfield Hodges, DPhil (Oxon), medical historian, co-editor of *Male and Female Circumcision: Medical, Legal, and Ethical Considerations in Pediatric Practice, Understanding Circumcision: A Multi-Disciplinary Approach to a Multi-Dimensional Problem,* and *Flesh and Blood: Perspectives on the Problem of Circumcision in Contemporary Society,* and co-author of *What Your Doctor May Not Tell You About Circumcision: Untold Facts on America's Most Widely Performed – and Most Unnecessary – Surgery,* and the author of many scholarly articles. Berkeley, California, USA.

Yngve Hofvander, MD, pediatrician and Professor, International Child Health, Uppsala University, Sweden, consultant to the Swedish International Development Authority, the World Health Organization, and UNICEF. Uppsala, Sweden.

Abdulcadir Omar Hussen, Gynecologist, Research Centre for Preventing and Curing Female Genital Mutilation and its Complications. Department of Gynecology and Obstetrics. University of Florence, Italy.

Anna Iordanidou, Graduate in Psychology, University of Padua, Department of Equal Opportunities of the Ministry of Interior (KETHI). Athens, Greece.

Stella Lineri, Padua Working Group on Female Genital Mutilation, Department of General Psychology, University of Padua, Italy.

Marica Livio, Ph.D. Faculty of Psychology, University of Padua, Italy

David J. Llewellyn, BA, JD, attorney at law, representing plaintiffs in wrongful circumcision and other genital injury cases. Atlanta, Georgia, USA.

Marilyn Fayre Milos, RN, co-founder and Executive Director of the National Organization of Circumcision Information Resource Centers (NOCIRC), coordinator, International Symposia on Circumcision, Genital Integrity, and Human Rights, editor, *NOCIRC Annual Newsletter*, and co-editor of *Sexual Mutilations: A Human Tragedy, Male and Female Circumcision: Medical, Legal, and Ethical Considerations in Pediatric Practice, Understanding Circumcision: A Multi-Disciplinary Approach to a Multi-Dimensional Problem,* and *Flesh and Blood: Perspectives on the Problem of Circumcision in Contemporary Society.* San Anselmo, California, USA.

Fabiola Pagani, Padua Working Group of Female Genital Mutilation, Department of Psychology, University of Padua, Italy.

Marianna Pappalardo, graduated in Psychology at University of Padua, Padua Working Group on Female Genital Mutilation, Department of General Psychology, University of Padua, Italy.

Claudia Piccolantonio, studied psychology at the University of Padua, now studying medicine at Johann-Wolfgang Goethe University in Frankfurt (am Main), focusing on women's rights and female genital mutilation, and working with Terre de Femmes. Frankfurt (am Main), Germany.

Chiara Rauhe, Padua Working Group on Female Genital Mutilation, Department of General Psychology, University of Padua, Italy.

Gritt Richter, Consultant on female genital mutilation, Terre des Femmes. Tübingen, Germany.

Lisa Rivaroli, Padua Working Group on Female Genital Mutilation, Department of General Psychology, University of Padua, Italy.

Petra Schnuell, studying ethnology, sociology, journalism, and communications at Georg-August University, Göttingen, and, since 1996, working with Terre des Femmes, especially with readings, seminars, and publications about female genital mutilation. Frankfurt (am Main), Germany.

Sara Simone, Department of Psychology, University of Florence Research Centre for Preventing and Curing Female Genital Mutilation and its Complications. Florence, Italy.

Saulo Sirigatti, Department of Psychology, University of Florence, Italy.

J. Steven Svoboda, JD, graduated cum laude from Harvard Law School, founder and Executive Director of Attorneys for the Rights of the Child (ARC), a non-profit organization addressing the illegality of involuntary genital surgery. Berkeley, California, USA.

Eleanora Tita, Padua Working Group on Female Genital Mutilation, Department of Psychology, University of Padua, Italy.

Franco Viviani, PhD, graduated in biological sciences from the University of Padua, Professor of Psychobiology at both the Faculty of Medicine and of Psychology, University of Padua, director of the 1986 film, *Somali Schoolgirls Speak*, a documentary on female genital mutilation in Somalia, participant in a field research project on genital mutilation and, that same year (1991), along with Professor Pia Grassivaro Gallo, received honorable mention from the Royal Academy of Overseas Science (Belgium). He has organized surveys, research projects, and symposia on female genital mutilation. Padua, Italy.

Elisabetta Villa, Padua Working Group of Female Genital Mutilation, Department of Psychology, University of Padua, Italy.

Hugh Young, BSc, graduated in zoology from Canterbury University, worked as a journalist and radio producer, and, from 1994 to 1995, reviewed the literature on safe sex for the New Zealand AIDS Foundation. He created and maintains a human rights website, www.circumstitions.com, a resource of references, quotations, arguments and visuals, particularly for intact males and their parents, treating intactness rather than circumcision as normal. He has an interest in linguistics and lexicography (as the website's glossary shows) and coined the word "circumstition." Pukerua Bay, New Zealand.

Chapter 1

CIRCUMCISION AS A MEMEPLEX

Hugh Young

Abstract: The real reason or reasons for circumcision remain mysterious. Part of the explanation is that circumcision is a memeplex, a cluster of related and interconnecting cultural units that are transmitted by imitation (memes), as well as by language. This paper considers the interrelationship and transmission of the constituent memes of circumcision, especially that of male infants, in the context of sexual selection and altruism.

The term "meme" was coined by Richard Dawkins, who defined it as "a unit of cultural transmission, or a unit of *imitation*."[1] The Oxford English Dictionary defines it as "an element of a culture that may be considered to be passed on by non-genetic means, esp. imitation."[2] Susan Blackmore says it is "whatever it is that is passed on by imitation."[3] Memetics is the study of memes. Common examples of memes are "tunes, ideas, catch-phrases, clothes fashions, ways of making pots or of building arches."[4]

Unlike genes, only human, and possibly some other primate, and electronic brains carry memes. Like genes, memes endure and copy themselves with variations, which are naturally selected, and thereby evolve. Like genes, they are "selfish," or rather, their function is the promotion of their own survival.

> ...all that counts in the life of a meme is whether or not it survives and replicates. ... we must remember that [saying that memes 'want,' 'need,' or 'try to do' something] is only shorthand for saying that the 'something' will improve the chances of the meme's being copied. Memes do not have conscious intentions; nor do they actually strive to do anything at all. They are simply (by definition) capable of being copied, and all their apparent striving and intentionality comes from this.[5]

G.C. Denniston et al. (eds.), Bodily Integrity and the Politics of Circumcision, 1—16.

A good analogy for a meme today is a computer virus. In fact, a computer virus, being merely information stored in, and transmitted between, information processing systems, is more like a meme than it is like a biological virus, which is a self-contained physical entity that actually moves between organisms, carrying genetic information with it.

Richard Dawkins, in his preface to Blackmore's book, writes:

> As with genes, we track memes through populations by their phenotypes [ways of being physically expressed]. ...the phenotypes of genes are normally parts of living bodies. Meme phenotypes seldom are.
>
> But it can happen. ...[A] Martian geneticist, visiting [my] school during the morning cold bath ritual, would have unhesitatingly diagnosed an 'obvious' genetic polymorphism. About 50 percent of the boys were circumcised and 50 percent were not. ...It is, of course, not a genetic but a memetic polymorphism. But the Martian's mistake is completely understandable; the morphological discontinuity is of exactly the kind that one normally expects to find produced by genes.
>
> [T]hrough most of history circumcision has been longitudinally transmitted as a badge of religion (of parents' religion I hasten to point out, for the unfortunate child is normally too young to *know* his own religious mind). Where circumcision is religiously or traditionally based (the barbaric custom of female circumcision always is), the transmission will follow a longitudinal pattern of heredity, very similar to the pattern for true genetic transmission, and often persisting for many generations. Our Martian geneticist would have to work quite hard to discover that no genes are involved in the genesis of the roundhead [circumcised] phenotype.[6]

Dawkins errs in saying only *female* genital mutilation is "barbaric" or that it is always religiously or traditionally based: in Egypt, FGM is defended on medical as well as traditional grounds.

> Dawkins says that a meme survives and spreads if it has "longevity, fecundity, and copying-fidelity."[7]
>
> Longevity: the individual circumcision lasts a lifetime (*pace* foreskin restoration), and the custom, as its proponents are wont to remind us, has persisted for thousands of years.
>
> Copying-fidelity: within remarkably broad parameters, all circumcisions are visually alike, constrained by the parameters of penile anatomy.

> Fecundity: the age, circumstances, and especially reasons given for circumcision, are subject to evolutionary change, and this paper focuses on these factors.

A memeplex is a cluster of related memes. Dawkins called it a "co-adapted meme-complex," but the name of the meme has evolved. Examples of memeplexes are religions, clusters of customs such as Christmas celebrations, and languages. Defining the boundary between a meme and a memeplex is a problem, but circumcision defined minimally as cutting the foreskin off the penis is clearly a meme, while the full context giving rise to circumcision is clearly a memeplex.

In the context of those memes, such as sexual taboos that promote having more children to whom the taboos are taught, thus causing the memes to proliferate, Blackmore writes:

> Given that young men have a strong desire for sex, dissuading them from masturbation is likely to increase the amount of vaginal sex they will have, thereby increasing the number of the offspring to whom they can pass on the taboo. (Lynch, 1966),[8] Lynch suggests a similar explanation for the success of the circumcision meme, because circumcision makes masturbation more difficult, but not vaginal sex.[9]

However, there is no evidence that circumcision significantly increases the birthrate, and many men claim their circumcision does not make masturbation more difficult, but the belief that it did was sufficient to establish circumcision in the United Kingdom and then the United States, late in the nineteenth century, and once established, it sustained and sustains itself by associated memes that had and have nothing to do with masturbation or sex. In fact, for most of the twentieth century, any role of the foreskin in sex has been ignored or denied in circumcising cultures.[10]

Immerman hypothesizes that circumcision "reorganizes the male's sensory somatocortex to raise the threshold of sexual excitability/distraction" making young men "slightly more tractable" and "slightly more restrained sexually and more cooperative in the pair bond."[11] These are debatable as evolutionary/biological goals for young men, and societies in which these are deemed desirable do not necessarily claim them as purposes for circumcision.

Helsten says "All societies have found the arguments that best fit their local cultural traditions and environments in order to introduce or maintain genital mutilation in its various forms."[12]

While Dawkins and Blackmore consider circumcision a meme, the central idea of cutting part of a baby's or child's penis off is always embedded in culture, tradition, religion, and/or medicine. That is why it is so dangerous to draw any conclusions from correlations between circumcision and any of the other accoutrements of a society, such as the incidence of some disease. There are always other factors to consider. While hundreds of distinct reasons for circumcision have been offered,[13] they fall into a few broad classes.

Here are some of the main ideas around circumcision:

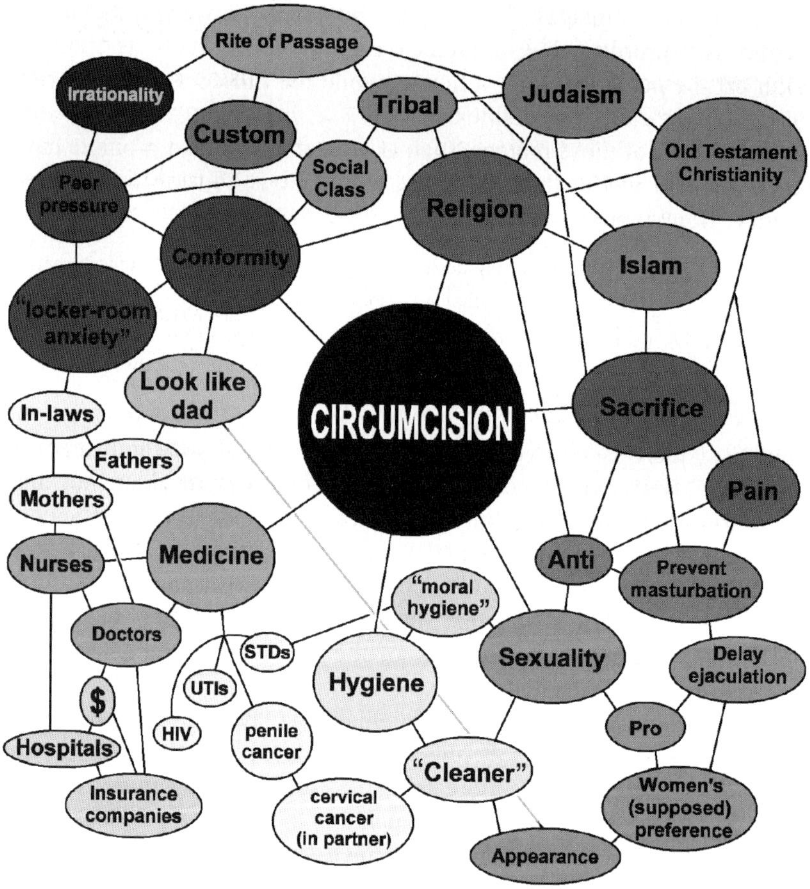

1. THE CIRCUMCISION MEMEPLEX

Things are actually more complicated than that. Ideas from one side of the diagram are frequently entangled with those from the other. For example,

> Masturbation is an individual act, and suppressing it helps maintain conformity.

The role of the father and perhaps the doctor is entangled with that of Old Testament patriarchs.

US women who prefer circumcised penises do so because custom has made that the only kind they encounter.

Women's supposed preference for the appearance of a circumcised penis connects with the father's wish that the son look like him.

In spite of such cross-connections, ritual circumcision and routine infant circumcision, i.e., without medical indication, might almost be considered different memes that happen to share the same phenotype:

Muslim ritual circumcision, done in infancy or childhood, is associated with ideas of ritual cleanliness, conformity, and as a rite of passage to manhood, as well as the belief that it is required by Islam. Customs vary across the huge Islamic world, but it is usually associated with celebration, feasting, and treating the boy as "Prince for a day," all parts of the Muslim memeplex that cement it into that culture. We have one report that Muslim circumcision is surgically much milder than western "medical" circumcision,[14] and this could explain how it can be done to boys old enough to speak.

Jewish ritual circumcision takes place at a social occasion for adults where the emphasis is on food. It is common for a meal, celebrating a symbolic act to take center stage and eventually replace the act. Much later, its subjects are explicitly taught about its significance (the meme is transmitted very deliberately) although that significance itself is a whole complex of ideas, some unrelated:

To keep a bargain Abraham made with God

As a badge of Jewish identity

To make him look different from outsiders

To remind him of anti-Semitic persecution

To bond him with his community

To educate him in his parent's faith

"... because Jewish men should be able to feel the pain of others more easily."

To symbolize humanity's unique essence as more than animal

To offer our children to a higher spiritual life

To ensure a share for him in the world to come

To draw down the Divine light, bring down the soul of holiness into the body, reveal the Jew's inherent connection to God

To signify the union of body and spirit

Because "[t]his paring away of the superfluous skin allows for Shechinah energy (the 'essence' of the Universal Deity) to permeate the seed of Israel."[15]

Because the foreskin concentrates negative energy

To "spiritually remov[e] and eliminat[e] undesirable character traits...depressive tendencies and so on..."

[To] eliminate from the body of the child, forces which might try to cultivate overindulgence in physical pleasures, etc."

For "its positive effects on the generations to come."

These might be summarized as "Identity," although it is generally agreed that circumcision is not what makes a boy Jewish, and supposed spiritual benefits, and the good feeling that some Jewish men have, of continuity with their age-old tradition through being circumcised, cannot be gainsaid. Yet ritual cleanliness is the invisible guest at the feast. Jewish defenders of circumcision frequently cite cleanliness and express horror at the supposed uncleanliness of the foreskin, as in this novel, set in sixteenth-century Portugal:

> Opening the draw string of his pants, he allowed the filth of his uncircumcised penis to unsheathe into the air.[16]

2. ROUTINE INFANT CIRCUMCISION IN THE USA PERPETUATED BY STEALTH

Circumcision is usually done out of the parents' sight, and at one time they were not even told it was to be done. Two parents might agree to circumcise their son for reasons that have nothing in common, might even be contradictory, and never share their reasons.

Its victims are discouraged from discussing it or even thinking about it. This silence has a component that is peculiar to circumcision and in part arises from the taboo on sex. Thus the transmission of the meme is often done indirectly, through euphemism ("a little snip") and with a kind of desperate jocularity ("I'm just taking him away for his circ.").

The ostensible reasons are:

medical, hygiene, and cleanliness

supposed psychological benefits of looking like the father, coupled with tradition or custom

sexual, coupled with women's supposed preference

fear of difference, projected into the child's future locker-room experiences

Each of these is a cluster of ostensible reasons, sometimes contradictory, for example, to both increase *and* decrease penile size or sensitivity.

The memes of Jewish circumcision and routine infant circumcision in the US are like the two members of a double star orbiting each other, influencing each other while keeping their distance. The interaction can be seen in many TV sitcoms,[17] where the circumcision being discussed is ostensibly Jewish, but stripped of any religious connotations, and many of the other "reasons" are adduced. An unknown number of US Jewish babies are reportedly circumcised in hospital before the eighth day, nullifying any covenantal value, but satisfying other memes of the memeplex.

The most effective memes are simple and couple the meme with something universally agreed to be good:

God said we must circumcise

All the nicest people are circumcised

Circumcision is American

Circumcised penises are cleaner

Circumcision protects against AIDS (or cancer, or whatever is the most feared disease of the day)

These associations make them hard to argue. Meme theorists call these associated ideas "baits" (promises to reward the host) and "hooks" (the part of a meme that urges replication).[18]

Two important memes — both are of the form "circumcise for the sake of someone else" — are complementary, but not exactly so:

3. "A BOY SHOULD LOOK LIKE HIS FATHER" AND "WOMEN PREFER CIRCUMCISED MEN"

"A boy should look like his father" seems to be spread mainly by *women.* In Western society, it sounds like popular wisdom and it can be spread without mentioning circumcision directly. It purports to be about the son's feelings, coupled with "locker-room anxiety," but it is really about the father's feelings: it is very rare for a boy to wish or ask to be circumcised to look like his father.

At a deeper level, it links to the more general "looking like his father" that assures the father that his wife's son is indeed his child. For that reason, it may be that the meme "a boy should look like his father" has a basis in biology:

Species that survive do so because they behave in ways that promote the survival of their own genes.

Individuals promote the survival of their own offspring and, in the wild, have no interest in the offspring of unrelated others or even a negative interest, as where a conquering alpha male lion kills the offspring of his predecessor.

Appearance is one of the ways individuals recognize their own offspring.

A male may be less likely to bond with his son if the appearance of his son's genitals is markedly different from his own. This fails to explain how he bonds with his daughters, but that may have a quite different mechanism, since the bond is different.

The boy's mother may fear that the father will not bond if their appearance is not similar and collude with making them match.

The question arises, how could circumcision get established if intact fathers risked failing to bond to their circumcised sons? The answer is that, in both the mythical origins of the religious rite, and probably its actual origins, and the historical origins of the surgical rite, such as Kellogg,[19] circumcision of boys out of infancy was customary for a generation or more before that of infants was, so intact fathers had intact sons, who grew up, were circumcised, and became the circumcised fathers of sons they then had circumcised. So there was no time when many intact fathers had circumcised sons.

How, then, did circumcision come to an end in Britain and New Zealand[20] without demur from the circumcised fathers of intact sons? A partial answer is that fathers and mothers were not consulted but were presented with a *fait accompli*, if doing nothing can be called a *fait*. The meme was never given the opportunity to be invoked.

This meme of "looking like his father" is related to circumcision as *a condition of, and rite of passage to, manhood*, the general pattern of tribal, as distinct from medical or religious, circumcision. A boy not only should look like his father, but actually must be made to look like his father, in order to become equal to him. It is notable that the three groups of cultures where this is the pattern, in Africa, Australia, and Polynesia, all separate boys from not only their mothers but also their biological fathers for the rite, and deliberately bond them with their peers instead, under the tutelage of the adult males collectively. On the Polynesian island of Tikopia,[21] the boy is cut by the most skilled man available, supported by his maternal uncle.

The complimentary meme, "Women prefer circumcised men" seems to be spread mainly by *men*. After all, women don't need to be told what women prefer.

It may be that in circumcising cultures women *will* prefer circumcised men, following the principle of "runaway sexual selection:" females prefer some characteristic, such as long tail feathers in males, even though or even because it puts the males at a disadvantage. If a female tries to buck the trend and mate with a male with short tail feathers, her short-tailed male offspring are less preferred by the next generation and her preference dies out.

In the same way, it makes genetic sense for women to prefer the kind of penis they think most women prefer. This has been demonstrated many times for characteristics indicating health, fertility, and the ability to provide for young. It may be the underlying reason for the expressed preference of US women for circumcised men.

The Williamsons' study of women in Iowa shows that the meme is widely believed by women where the great majority of men are circumcised even by the women who have no basis for comparison.[22] Women who do themselves prefer circumcised penises are of course more likely to believe and transmit the meme, while the others are silent, giving the impression that all do.

The meme "Women prefer circumcised penises" is likely to spread among men wherever a majority of men are circumcised, whether there is any truth to it or not because it comforts men to think they are of a preferred type.

This meme is extremely strong where circumcision is tribal, and hence universal within the tribe, taking the extreme form "No woman will look at a[n intact] man." Women in bars in Samoa and *fa'afafine* transsexuals, more or less, are said to check European men out manually before agreeing to go with them.[23]

When women hear the meme, they are likely to assume that men are speaking from experience, not just wishful thinking. If they are smart, they will not hurt circumcised men's vanity by telling them that they have no preference or actually prefer intact penises. So men's belief in the meme is reinforced and not challenged.

These two complementary memes mean that circumcision becomes a marker of conformity having nothing to do with hygiene, sexuality, religion or anything else, only itself. A circumcised penis proves that you have orthodox, "normal" parents.

There is no corresponding meme "Women prefer intact penises" in communities where most penises are intact, not because women don't prefer them there, but because, where intactness is already the norm, it is spread genetically, not memetically. Intact penises don't need memes to continue, only genes. Circumcision always needs memes. Intactness does need memes to supplant circumcision, to move into areas where circumcision is the norm, and genital integrity activism may be considered a memeplex.

4. THE MEMETICS OF FEMALE GENITAL MUTILATION

The meme "A girl should look like her mother," corresponding to the one for boys above, is not greatly recorded in cultures that practice FGM, if at all.

"The motivation for the practice varies from setting to setting and reflects beliefs and cultural mores that include religious, health, and social factors. For example, FGM is believed to maintain cleanliness, increase a girl's chances of marriage, protect her virginity, discourage "female promiscuity," improve fertility, prevent stillbirth."[24]

This is probably because questions of maternity seldom arise. The very similar idea, "I want her to look like me," has been heard,[25] but the difference is that it does not pretend to be altruistic.

The complementary meme, "Men prefer circumcised women," is very strong in cultures that practice FGM. In fact, it takes the extreme form "She *must* be circumcised or *no* man will want to marry her." This is probably spread more by women, since the marriage of their daughters and continuation of their genes, including the extra-nuclear, mitochondrial gene line, is of vital importance to them.

Men collaborate, not by preferring circumcised girls as such, although in a tribal society, her genital status will not be a secret, but by refusing to take a woman who has been rejected by all other men, since in most other circumstances, this is a marker that she is somehow genetically unfit.

FGM is also spread by the meme "FGM makes a woman faithful" and, to the extent that it makes intercourse difficult or impossible without defibulation or opening and makes it unpleasant then, it has some truth. The biological function of that is, as always, that it reassures a husband that it is his genes he is investing energy in, when his partner gives birth.

5. INTERSEXUALITY

When a child is born intersexed, their parents and their deliverers are thrown into a panic because the idea "Humans are male or female" is immensely strong, so strong that it seldom needs to be expressed in so many words, but is built into many of our languages through grammatical gender, being intimately associated with our genes. A meme for preferring intersexuality would have no future. What is striking is that the surgical action that is taken in response to that panic has little or no bearing on the child's genes, or their reproductive future, but is concerned almost entirely with the appearance of their genitals. Regardless of biological facts, the ultimate test for humans of another person's sex is the appearance of their genitals, and the meme "A child should look like its parents" kicks in for both parents.

6. THE ALTRUISM MEME

Blackmore illustrates that altruism is a powerful way of spreading memes. Hence, many memes spread by linking themselves to altruism.

> People are nice to each other to get kindness in return, and their emotions are designed appropriately. ...Kind and generous behaviors will spread by imitation, ...behaviors that *look* like kind and generous ones, or are prevalent in kind and generous people, will also be spread by imitation.
>
> ...if you are in a community that uses reciprocal altruism, you are likely to gain most by being with people who are known to be generous. So the generous people will have more contact with others and therefore more opportunities for spreading their memes.[26]
>
> Memes that have nothing to do with altruism can benefit from "copy-the-altruist" by just tagging along for free. ...we can expect memes to have devised strategies for getting into altruistic people without actually being altruism memes themselves (or more accurately, memes that happen to have such strategies should have survived better than those without, and we should be able to observe them around us). Are there such examples? Yes. They range from

> little groups of co-memes to very complicated memeplexes. ...the essence of any memeplex is that the memes inside it can replicate better as part of the group than they can on their own.[27]

Circumcision is an excellent example of such a memeplex, the cluster of ostensible reasons for doing it is hydra-headed and ever-changing. The silly ones are carried on with it, as well as those with any validity at all.

The ostensible reasons for infant circumcision always have an altruistic component. Even the reason "to punish him for masturbation" was only prevalent when that was believed to be for his ultimate betterment, including the salvation of his immortal soul.

Blackmore cites Allison[28] on "beneficent norms" applied to religion:

> If you see someone else who acts the same way as you do, it is likely that you both have cultural ancestors in common. If you now help him, you make it more likely that he will be successful, and hence that he will pass on his memes, including the rule 'Be good to those who act like you.' Allison calls this a 'marker scheme.' He gives the examples of wearing a turban or abstaining from certain foods. ...He adds that markers that are costly or difficult to learn can deter exploitation by outsiders. Apart from languages, a good example is religious rituals. Many of these require years to learn and others, such as ritual circumcision, are certainly costly for an adult.[29]

7. SO COSTLY THAT RITUAL CIRCUMCISION OF CONSENTING ADULTS IS VERY RARE AND EVEN SOME BRANCHES OF JUDAISM ALLOW CONVERTS TO FORGO IT

Unlike those beneficent norms, religious, ritual, and "medical" circumcision and FGM are forced on infants and children by adults who have already undergone it themselves. The ones who perform it are not those who suffer now. Their culture or religion has taught them that it is "inevitable" and "necessary" and "beneficial" so that they are able to suppress their natural revulsion, and even wallow in their own empathy with the child's pain. They have usually forgotten their own pain and may take comfort that their sons will too. If they remember the pain, they may consider that it, too, is valuable, as Nelson Mandela does.[30] So by a variety of dodges, the circumcision meme can pass itself off as altruistic. Or rather, people infected with it can.

8. THE SPANNER IN THE WORKS

Meanwhile, the irrational aspect to the circumcision memeplex adds a complication. Circumcising a baby is "doing something" at a particular time and place, so it has a definiteness about it that leaving the baby alone does not. In fact, US mothers who leave their babies intact are frequently accused of neglect.

Circumcision leaves a vivid mark of having been done. It does not usually do sufficient harm at any one time to be genetically or socially contraindicated, unlike, say, castration. There is some suggestion that ritual circumcision may have been introduced in Egypt as a substitute for ritual castration of a priestly caste rather than the whole male population, obviously. The good that circumcision supposedly does is set in the unforeseeable future. In this, circumcision is like those religion memes that promise infinite and eternal rewards or punishments after death. Since the evils it is supposed to prevent are rare, it is allowed to take credit by default for their absence. In this, it is something like a lucky charm or wearing garlic to ward off vampires.

Like genes, memes evolve to maximize their transmission. Secular circumcision quickly evolved from being a childhood treatment for masturbation to an infant prophylaxis for it, largely because newborns can put up less resistance. Jewish circumcision probably moved in the same direction in ancient times. Muslim, Korean, Philippine, and tribal circumcision have yet to do so, though there are some trends in that direction, concurrent with the medicalization of the ritual. When masturbation hysteria waned, circumcision, or rather, people's wish to circumcise — for that is the heart of the circumcision memeplex — attached and reattached itself to the successive diseases it was alleged to prevent. The concept of a meme or memeplex is particularly productive with regard to genital mutilation because the idea of genital mutilation has a life of its own that is independent of individual rationality.

This paper will not go into circumcision for the sexual gratification of the circumciser, because its extent and role is unknown and unstudied, but it certainly exists.[31]

9. INTACTNESS, THEN, IS ALSO A MEME?

An intact penis or female genitalia is not itself a meme nor a meme-phenotype because it is not a behavior, and it is transmitted by genes, not by imitation. In the context of a strongly circumcising culture, it may require unusual determination to break the hold of the circumcision meme, but, for the most part, leaving a baby alone needs no reason.

However, the campaign for genital integrity, which only exists in the context of genitally mutilating cultures, may usefully be considered a meme or memeplex. Here are some of the ideas linked to it:

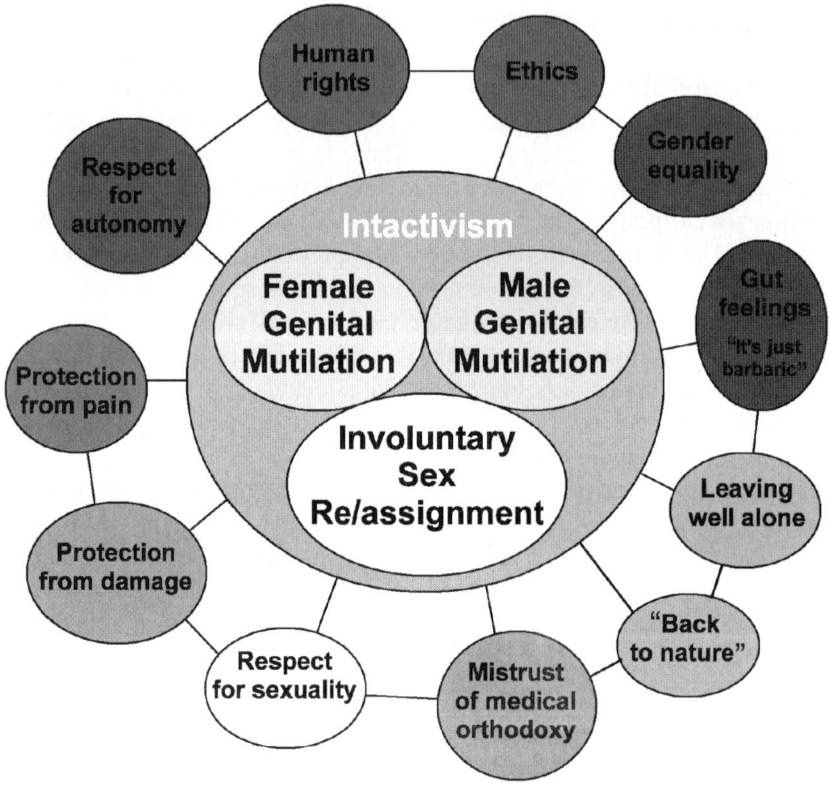

10. THE INTACTIVISM MEMEPLEX

Opposition to the three varieties of genital modification has been amalgamated here because most of the associated factors relate to all three in greater or lesser degree. I put the "gender equality" meme near to male circumcision, but there is a special relationship between feminism and opposition to FGM — women's outrage that this can be done to women. We might speak of "gender solidarity." There seems little or no equivalent men's outrage, based on gender solidarity, that this can be done to fellow men. It seems to be overwhelmed by a macho denial of pain imposed on tiny babies.

The genital integrity activism memeplex is clearly much simpler than the circumcision memeplex. The different elements also reinforce each other in ways that the parts of the circumcision memeplex do not. These are two of the strengths by which the genital integrity activism memeplex may ultimately prevail. Genital integrity activism is an easy idea to transmit, and once transmitted, it is not easily lost.

REFERENCES

[1] Dawkins R. Memes: the new replicators. In: Dawkins R. The Selfish Gene. Oxford: Oxford University Press, 1989.

[2] Little W, *et al.* The Oxford English Dictionary. Oxford: Oxford University Press, http://dictionary.oed.com/.

[3] Blackmore S. The Meme Machine. Oxford: Oxford University Press, 1999. p. 43.

[4] Dawkins R. Memes: the new replicators. In: Dawkins R. The Selfish Gene. Oxford: Oxford University Press, 1989.

[5] Blackmore S. The Meme Machine. Oxford: Oxford University Press, 1999. p. 162.

[6] Dawkins R. Preface in: Blackmore S. The Meme Machine. Oxford: Oxford University Press, 1999. p. xiii.

[7] Dawkins R. Memes: the new replicators. In: Dawkins R. The Selfish Gene. Oxford: Oxford University Press, 1989.

[8] Lynch A. Thought Contagion: How Belief Spreads Through Society. New York: Basic Books, 1996.

[9] Blackmore S. The Meme Machine. Oxford: Oxford University Press, 1999. p. 135.

[10] Masters WH, Johnson VE. Human Sexual Response. Boston: Bantam, 1966. pp. 189-190.

[11] Immerman RS. A proposed relationship between circumcision and neural reorganization. Journal of Genetic Psychology 1998;159(3):367-378.

[12] Hellsten SK. Rationalising circumcision: from tradition to fashion, from public health to individual freedom — critical notes on cultural persistence of the practice of genital mutilation J Med Ethics 2004;30:248-253.

[13] Young H. More than 330 Circumstitions. http://www.circumstitions.com/Stitions&refs.html.

[14] Amber Craig wrote: "Baba Jallow [came to] speak ... about his new book *Dying for My Daughter*. ... he remembered his circumcision [in Gambia]... he and a bunch of pre-adolescent boys were "taken to the bush" and a slit was made in the frenulum." "Baba understood [circumcision] to be cutting the frenulum and could not understand our concerns. Now he does." — Marilyn Milos.

[15] "Yechidah," 12 June 2004 (Shechinah energy is defined by Riain Barton as "the 'essence' of the Universal Deity").

[16] Zimler R. The Last Kabbalist of Lisbon. London: Arcadia, 1999. p. 54.

[17] Young H. TV Sitcoms. http://www.circumstitions.com/TVSitcoms.html.

[18] McFadzean D, Grant G, *et al.* The Virian Lexicon http://virus.lucifer.com/lexicon.html.

[19] Kellogg JH. Plain Facts for Old and Young. Burlington, Iowa: I.F. Segner, 1882.

[20] McGrath K, Young H. A review of circumcision in New Zealand. In: Denniston GC, Hodges FM, Milos FM (eds). Understanding Circumcision. New York: Kluwer Academic/Plenum, 2001. pp. 129-146.

[21] Firth R. We the Tikopia: Kinship in Primitive Polynesia. Boston: Beacon Press, 1957. p. 406.
[22] Williamson ML, Williamson PS. Women's preference for penile circumcision in sexual partners. Journal of Sex Education and Therapy 1988;14(2):8-12.
[23] McGrath K. personal communication.
[24] International Planned Parenthood website. [cited 2004 July 20] http://www.ippf.org/medical/imap/ statements/eng/2001_5c.htm.
[25] Burney T. Giving treatment, but not stirring shame. The New York Times (5 June 2004) http://www.nytimes.com/2004/06/06/health/womenshealth/06BURN.html?ex=1087883608&ei=1&en=ae8d7d8a5cf9a4a3.
[26] Blackmore S. The Meme Machine. Oxford: Oxford University Press, 1999. p. 156.
[27] Blackmore S. The Meme Machine. Oxford: Oxford University Press, 1999. p. 168.
[28] Allison PD. The cultural evolution of beneficent norms. Social Forces 1992;71:279-301.
[29] Blackmore S. The Meme Machine. Oxford: Oxford University Press, 1999. p. 191.
[30] Mandela N. Long Walk to Freedom. London: Little Brown; 1994. pp. 510-511.
[31] Circlist. Undated [cited 1 August 2004]: http://www.circlist.com/.

Chapter 2

THE LIFE OF THE FLESH IS IN THE BLOOD

The Meaning of Bloodshed in Ritual Circumcision

Leonard B. Glick
Professor of Anthropology Emeritus, Hampshire College, Amherst, MA, USA

Abstract: The central rite in Temple Judaism was animal sacrifice. With the destruction of the temple, the only remaining candidate for a blood sacrifice was circumcision. The circumcision rite began with emphasis on the idea of covenant; the rabbis retained this but superimposed sacrifice of the infant's genital blood as a redemptive rite. Circumcision still meant foreskin removal, but now it also meant ritual bloodshed.

INTRODUCTION: "A NECESSARY PART"

In a December 2000 letter to a Toronto newspaper, a physician-mohel named Aaron Jesin objected to having been misquoted in an article on circumcision published several days earlier: "I did not say that ritual circumcision causes no bleeding," he wrote. "Some amount of bleeding is a necessary part of the ritual circumcision. To suggest that I do them without any bleeding is damaging to my reputation in the Jewish community."[1]

Jesin was correct; bloodshed is indeed an essential feature of every ritual circumcision. Although some ritual circumcisions nowadays are accompanied by comforting homilies and cheerful banter intended to distract attention from the physical reality of the proceedings, the authentic rite, as conducted by Orthodox mohels, accords full recognition to the fact that the infant's genitals have been made to bleed. Yes, foreskins must be removed, but even when there is no foreskin, blood still must flow. A prospective male convert to Judaism who is already circumcised must nevertheless undergo a ritual circumcision, called *hatafat dam brit*, "shedding the blood of the covenant," in which a drop of blood is drawn from his foreskin remnant.[2]

G.C. Denniston et al. (eds.), Bodily Integrity and the Politics of Circumcision, 17 –36.

But consider this: the foundational text for Jewish circumcision, Chapter 17 of Genesis, does not mention blood. And although it might be said that sacrificial bloodshed is implied when a foreskin is removed, the emphasis of the chapter is clearly not on sacrifice but on *covenant*. The Lord tells Abraham how he must seal the covenant that will endow him with abundant progeny and a huge land grant: "you shall circumcise the flesh of your foreskin," he declares, and "throughout the generations, every male among you shall be circumcised at the age of eight days." Nor is there reference to blood in the brief injunction in Chapter 12 of Leviticus, which simply repeats the same message: "on the eighth day the flesh of his foreskin shall be circumcised."[3]

Several connected questions arise: If bleeding is essential in a ritual circumcision, why is there no mention of blood in the biblical texts? Is it possible that bloodshed was only incidental in the original version of ritual circumcision, but achieved greatly heightened significance later? If so, when did this happen, and why?

1. COVENANT

The belief in a covenant between the Lord and the Hebrews began long before anyone heard the story of Abraham's covenant *as described in Genesis 17*. To explain this, I have to say a few words about how the first part of the Hebrew Bible (the Torah, or "Five Books of Moses") was created. Although no one can be certain about details, biblical scholars now generally agree that the five books consist of four distinct segments, composed at various times and eventually combined around 500 BCE. They label the texts J, E, D, and P. For our purposes, those that matter are J, E, and P. J and E were written around 800 or 900 BCE; P, the final segment, was composed by a temple priest, or perhaps several priests, around 500 BCE, and was blended into the earlier texts.[4] The dates are obviously approximate; the significant point here is that J and E are substantially older than P.

The first time we learn about the Lord's covenant with Abraham comes not in Chapter 17 of Genesis but in Chapter 15. The story is nearly the same in most essentials (a promise of immense reproductive success and a generous land grant) —but with one crucial difference: Genesis 15 says nothing about circumcision. Only in Genesis 17 does Yahweh demand that Abraham seal the covenant with circumcision. Genesis 15 is a J text; Genesis 17 is a P text. In other words, the *original* story of Abraham's covenant antedates the version with circumcision by several hundred years —meaning that belief in covenant long preceded the connection between covenant and circumcision.

1.1 Blood of the Covenant

Most Jews nowadays associate the idea of covenant almost entirely with circumcision. But there was another early covenant narrative, far more comprehensive than Genesis 15, and more momentous for the development of Judaism: Yahweh's covenant delivered to Moses at Sinai, when the Israelites learned about the almost innumerable laws and regulations they would now be required to obey. The story of that covenant never mentions circumcision, but it does include a dramatic scene of blood sacrifice. The Sinai narrative is complex, since it consists of several texts composed at different times and later woven together; but for our immediate purpose the most significant passages are in Chapter 24 of Exodus. *This is an E text*, written, we recall, sometime around 800 or 900 BCE. In other words, although the *account* of the Sinai covenant necessarily appears in the biblical narrative sequence much later than the one about Abraham's circumcision, the actual *text* was composed centuries earlier. And what we learn from that text is that belief in blood sacrifice as a way of confirming a covenant long antedated the idea of a relationship between covenant and circumcision.

Here is the relevant part of the Sinai narrative. Moses first instructed the Israelites on "all the commands of the Lord." Then he erected an altar at the foot of Mount Sinai and assigned several young men to sacrifice bulls "as offerings of well-being to the Lord." Following the sacrifices he completed the ritual:

> Moses took one part of the blood and put it in basins, and the other part of the blood he dashed against the altar. Then he took the record [*sefer*, "book"] of the covenant and read it aloud to the people. And they said, "All that the Lord has spoken we will faithfully do [we will do and obey]." Moses took the blood and dashed it on the people and said, "This is the blood of the covenant which the Lord now makes with you concerning all these commands."[5]

Here, then, we have a biblical text, several centuries older than Genesis 17, which plainly unites blood and covenant in a single compelling image. Sprinkling blood on the altar was customary practice in temple sacrifices. The distinctive event here was splashing it on the assembled Israelites (perhaps only men, although the text says *am*, "people"). This was a rite of purification and consecration; the sacrificial blood sanctified the participants, rendering them fit for reception of the covenant.[6]

Aside from a single late exception, this is the only biblical passage containing the term "blood of the covenant." (The exception appears in the book of the prophet Zechariah, 9:11: "You for your part, have released your prisoners from the dry pit [i.e., a dungeon], for the sake of the blood of your covenant."[7] Zechariah lived in the sixth century BCE, but the dating of this passage is uncertain.)

The Sinai narrative tells us that sacrificial blood validated, or confirmed, the most consequential of all covenants. Why, then, was Abraham's covenant sealed only by foreskin removal? I think the answer lies in the nature of the religion instituted by the priests who composed Genesis 17 — the religious system we now call *Temple Judaism*.

2. TEMPLE JUDAISM: SACRIFICE AND CIRCUMCISION

I said that the P text, which includes Genesis 17, was composed by a priest or priests, sometime around 500 BCE (when they were also integrating all four texts into the Torah). Bent on establishing a male-centered theocracy devoted to the exclusive service of Yahweh, the priests instituted a tightly regulated society emphasizing repentance for sin (whether willfully committed or not) and avoidance of ritual pollution. And they claimed that they themselves, hereditary priests, were the only persons qualified to perform sacrificial rituals in the Jerusalem temple —rituals centering on blood as offerings to their ever-vigilant deity.

2.1 Offering Blood to Yahweh

The essential component of every temple ritual was animal sacrifice. Priests sacrificed cattle, sheep, goats, and fowl, splashing blood on the altar and sometimes even daubing it on their own faces and garments. Blood was the vital element, the substance that redeemed those who had transgressed, pacifying Yahweh and appeasing his terrifying fury. For example, in Leviticus 4:3-7 (a P text), the Lord, speaking through his intermediary Moses, provides instructions for sacrifice of "a bull without blemish" by a priest who has "incurred guilt" (albeit "unwittingly"), "so that blame falls upon the people":

> He shall bring the bull to the entrance of the Tent of Meeting, before the Lord, and lay his hand upon the head of the bull. The bull shall be slaughtered before the Lord, and the anointed priest shall take some of the bull's blood and bring it into the Tent of Meeting. The priest shall dip his finger in the blood, and sprinkle of the blood seven times before the Lord, in front of the curtain of the Shrine. The priest shall put some of the blood on the horns of the altar of aromatic incense, which is in the Tent of Meeting, before the Lord; and all the rest of the bull's blood he shall

pour out at the base of the altar of burnt offering, which is at the entrance of the Tent of Meeting.[8]

Twice the priest is instructed to display the blood "before the Lord," to emphasize that, although priests may eventually eat sacrificial beef, blood belongs solely to the Lord.

In another passage in Leviticus (also a P text), Yahweh makes it absolutely clear that *all* blood belongs to him alone: "And if anyone of the house of Israel or of the strangers who reside among them partakes of any blood," he warns, "I will set My face against the person who partakes of the blood, and I will cut him off from among his kin. For the life of the flesh [*basar*] is in the blood, and I have assigned it to you for making expiation for your lives upon the altar; it is the blood, as life, that effects expiation."[9]

2.2 Offering Foreskins to Yahweh

Although Genesis 17 contains the first mention of circumcision in the Bible, ritual circumcision began long after the supposed lifetime of the mythical Abraham (who is said to have lived around 1800 BCE).[10] That happened only around 500 BCE, when the temple priests redefined an ancient custom as a mandatory rite of initiation into their theocracy. Circumcision of male children (but probably not infants) had been practiced for centuries, by Israelites and their neighbors, as an offering of part of the child's penis — a gift to whatever deity reigned in his time and place — perhaps to protect the child from evil spirits or to redeem a first-born son from actual sacrifice. At times it may have been a father's personal sacrifice on his own behalf.[11] But circumcision was not required of anyone; the decision to circumcise was made by each father individually.

The Judean priests appropriated this practice, converting it into a mandatory rite to be undergone by *all* male infants on the eighth day of life. The child's father was to perform the operation himself, as a sign of personal loyalty and submission to priestly authority; in practice, he could delegate another man to do the actual cutting for him, but always in the father's presence and with his expressed consent. To validate the new rite, the priests introduced a second version of the Genesis 15 narrative, adding the circumcision command to Abraham as a requirement for the covenant. This became Chapter 17 of Genesis. Both versions were incorporated into the final redacted version of the Torah.[12] As we've seen, Genesis 17 says nothing about blood; what seals the covenant is foreskin removal.[13] Whereas it was the *privilege of priests* to offer bloody sacrifices to Yahweh, it was the *duty of fathers* to offer infant foreskins to him, as testimony of their loyalty to the patriarchal theocracy established by those priests, and of their promise to raise their sons as equally loyal successors. The rite thus accomplished two main purposes: it initiated a boy into the theocracy, while confirming his father's readiness to submit to priestly authority.

Of course, foreskin removal was also a kind of sacrificial offering, and blood loss as an accompaniment to that sacrifice must have been taken for granted. But if sacrificial blood was of such supreme importance to Yahweh, why did the priests (authors, we remember, of Genesis 17) focus exclusively on foreskins and say nothing about bloodshed? Why did their narrative not have Yahweh insist to Abraham that he and his descendants must "circumcise the flesh" of their foreskins *and* display the child's blood as a special offering? I think the answer is that they conceived and instituted circumcision and sacrifice for entirely different purposes. Animal sacrifice was an exclusively priestly act, effecting purification and release from punishment for sin. The purpose of circumcision — initiation into a community of covenanted males, performed by men who were not priests — differed so sharply from that of temple sacrifice that blood display might even have seemed improper.

But the situation changed, dramatically and forever, when the temple was destroyed and animal sacrifices became a thing of the irretrievable past. How rabbis — successors to the priests and creators of a new form of Judaism — dealt with that dilemma is our next subject.

3. RABBINIC JUDAISM: CIRCUMCISION AS BLOOD SACRIFICE

Temple Judaism, dependent on the existence of a clearly defined territorial community, ended in the late first century CE with the Roman conquest of Jerusalem, the destruction of the temple, and the dispersal of many of the remaining Judeans. I say "remaining" because in fact the so-called Diaspora had begun centuries earlier, as Judeans settled throughout the eastern Mediterranean region, particularly in such urban centers as Alexandria, Damascus, and Antioch. Living far from Jerusalem and able seldom if ever to travel there even for major festivals, they developed their own version of Judaism, focusing on local synagogues: religious centers where they met not only for prayer, but for study and all manner of social events. The leaders of local communities, men recognized and respected for their religious learning, were called rabbis, "masters" or "teachers," their status validated not by patrilineal descent but by personal knowledge and ability to interpret religious law. By the first century, rabbis had appeared as a source of authority in Judea as well. After 70 CE, when the destruction of the temple resulted inevitably in the disappearance of priestly functions, rabbis stepped in to take their place as communal leaders. In short, Temple

Judaism was succeeded by *Rabbinic Judaism*; and it is the latter that evolved into what we know as Judaism today.[14]

But although rabbis succeeded priests, it was never their intention to replace the religious system instituted by their predecessors. Rather, they sought to create a version of Judaism that preserved the traditional religion as much as possible. And although they obviously could not retain the sacrificial practices of Temple Judaism, they did their best to preserve the *memory* of those practices. This they accomplished partly by incorporating into the liturgy biblical texts describing sacrificial protocols supplemented by expressions of longing for a return to temple times.[15] But they also needed some kind of compensation for temple sacrifices — some ritual act or acts that would substitute, as it were, for what had been the principal form of homage to Yahweh. What that deity had most obviously demanded was blood, visibly displayed, his alone. With sacrificial animal blood no longer available, a substitute had to be found — a ritual performance featuring the all-essential blood.

What substitute was available? There remained only one ritual act that could be *reinterpreted and refashioned* as blood sacrifice: circumcision. To state the case in the bluntest terms, only one kind of ritual cutting of flesh survived; and although until then the inevitable bloodshed of that procedure had not come in for particular attention, now it moved to center stage. *The blood of circumcision became the sacrificial blood of the covenant.*

3.1 Two Redemptive Bloods: Paschal Lamb, Circumcised Infant

One biblical narrative in particular appears to have provided sanction and direction for the new definition of circumcision blood. The Passover narrative, recorded in Exodus, describes a sacrifice with unique purpose and significance. Whereas the blood displayed at temple sacrifices (mandated in Leviticus) served either to atone for sinful acts or to cleanse those who had become impure, the blood of the paschal lamb saved boys from death. Applied to doorposts and lintels, the blood was a redemptive sign; it notified Yahweh that he (or perhaps his surrogate, called the "Destroyer") should pass over Hebrew homes, sparing their first-born sons and slaying only those of the Egyptians. The text implies that unless alerted, Yahweh would kill all first-born boys. But the blood of the lamb, properly displayed, ensured that his murderous wrath would be directed only against the sons of Egyptians. "For when the Lord goes through to smite the Egyptians," Moses announces, "He will see the blood on the lintel and the two doorposts, and the Lord will pass over the door and not let the Destroyer enter and smite your home." And, he continues, when you enter Canaan, you must commemorate this event with an annual passover sacrifice; and when your children [or sons] ask what this means, you shall tell them, "It is the passover sacrifice to the Lord, because He passed over the houses of the Israelites in Egypt when He smote the Egyptians, but saved our houses [i.e., our sons]."[16]

The paschal sacrifice appears to have been originally a domestic ceremony, performed by pastoralists in the spring before leaving for summer pastures. The blood of a sacrificed lamb or kid was smeared on tents or doorposts, and sometimes on newborn animals as well. The purpose was to summon supernatural protection, not only for newborn animals but probably also for newborn children.[17] (It seems possible that blood was smeared on children as well, but that is speculative.) Perhaps as early as the seventh century BCE, the sacrifice was converted into a centralized ritual performed in Jerusalem for the entire community, leaving only a commemorative meal for household celebration, and that remained the case throughout the period of Temple Judaism, from the fifth century BCE onward.[18] By the second century BCE, domestic celebration of *Pesach* (Passover) may have become the most important ritual of the year, and for those living in the Diaspora it became the most popular time for pilgrimage to Jerusalem. At the temple celebration, the blood of paschal lambs was poured at the base of the altar, like other sacrificial blood, while the meat was ceremonially roasted and consumed at home.[19]

The Bible provides substantial evidence that in the minds of the Israelites, Passover and circumcision were linked — first, because according to the Exodus narrative, the blood of the paschal lamb redeemed male children; but also because, as we learn from the text, only circumcised men were permitted to share the paschal meal. The Lord dictates who may partake of future passover offerings. Slaves may share in the meal only after they have been circumcised. Similarly, resident aliens (men, of course) who want to join the community must first undergo the sanctifying operation: "He shall then be as a citizen of the country. But no uncircumcised person may eat of it."[20] (This seems to imply that no woman might participate under any circumstances.) Consumption of the paschal sacrifice was a ritualized communal act available only to properly sanctified males.

The rabbis discovered another connection between circumcision and the paschal sacrifice: both bloods were equally redemptive, a symbol (*ot*) of the covenant between God and the Israelites.[21] An early expression of this relationship appears in a third-century rabbinic text called *Mekhilta*. The discourse develops with reference to a narrative in Chapter 16 of Ezekiel, having nothing to do with circumcision or Passover originally, but selected for attention because of a double reference to blood. The Lord, speaking through Ezekiel, reprimands the backsliding Israelites, portraying them as an abandoned female infant whom he rescued: On the day of your birth, he declares, "you were left lying, rejected, in the open field. When I passed by you and saw you wallowing in your blood, I said to you: 'Live in spite of your blood' [literally, 'In your blood live']. Yea, I said to you: 'Live in spite of your blood.'" Later, he continues, "I passed by you [again] and saw that your time for love had arrived ... and I entered into a covenant with you by oath. ...thus you became Mine."[22] The author of *Mekhilta* (citing a predecessor) explains:

> This means, the time has arrived for the fulfillment of the oath which the Holy One, Praised be He, had sworn unto Abraham, to deliver his children. But as yet they had no religious duties to perform by which to merit redemption ...Therefore the Holy One, Praised be He, assigned them two duties, the blood of the paschal sacrifice and the blood of circumcision, which they should perform so as to be worthy of redemption. For thus it is said, "When I passed by you and saw you wallowing in your blood, I said to you, 'In thy blood live' and I said to you,'in thy blood live.'"[23]

The linkage between the two bloods appears in another rabbinic text, *Pirkei de-Rabbi Eliezer* ("Chapters of Rabbi Eliezer"), but here with explicit reference to the original paschal sacrifice in Egypt:

> The Israelites took the blood of the covenant of circumcision [and the blood of the Paschal lamb], and they put [them] upon the lintel of their houses, and when the Holy One, blessed be He, passed over to plague the Egyptians, He saw the blood of the covenant of circumcision upon the lintel of their houses and the blood of the Paschal lamb, He was filled with compassion on Israel, as it is said, "And when I passed by thee, and saw thee weltering in thy (twofold) blood, I said unto thee, In thy (twofold) blood, live; yea, I said unto thee, In thy (twofold) blood, live." "In thy blood" is not written here, but "in thy (twofold) blood," with twofold blood, the blood of the covenant of circumcision and the blood of the Paschal lamb; therefore it is said, "I said unto thee, In thy (twofold) blood, live; yea, I said unto thee, "In thy (twofold) blood, live."
>
> Rabbi Eliezer said: Why did the text say twice, "I said unto thee, In thy blood, live; yea, I said unto thee, In thy blood, live"? But the Holy One, blessed be He, said, "By the merit of the blood of the covenant of circumcision and the blood of the Paschal lamb ye shall be redeemed from Egypt, and by the merit of the covenant of circumcision and by the merit of the Passover in the future ye shall be redeemed at the end of the fourth kingdom"; and therefore it is said, "I said unto thee, In thy blood live; yea, I said unto thee, In thy blood, live."[24]

I have quoted this text in full, with its seemingly tedious (but essential) repetitions, to emphasize further how intimately the two kinds of blood now were thought to be connected. The early rabbis understood, of course, that sacrificial lamb blood was no longer available; what remained, though, was the blood shed at ritual circumcisions. These texts strongly suggest that, by equating the two bloods so explicitly, the rabbis were able to attribute newly heightened significance to circumcision blood. From then on, the doomed foreskin (always object for rabbinic expressions of contempt as a repulsive "blemish") was overshadowed for attention by male genital blood, the revered life-substance, with power to sanctify and redeem not only the infant but the entire people of Israel.

3.2 Flesh

"The life of the flesh is in the blood." Although this passage from Leviticus is often quoted in discussions of Judean blood sacrifices, the mention of "flesh" seldom receives notice. But for the student of the history of Jewish circumcision no word is more deserving of attention.[25] The Hebrew word *basar* ordinarily means "flesh" or "meat," and that seems to be its obvious meaning here. But in biblical Hebrew, specifically in the P (priestly) texts, the same word may also refer to the penis. That is clearly the case in Genesis 17; for example, in verse 11, the Lord commands Abraham to "circumcise the flesh [*basar*] of your foreskin," and in verse 13 he says, "Thus shall My covenant be marked in your flesh."[26] In Leviticus 15:2 (also a P text) Yahweh instructs Moses and Aaron in urology: "When any man has a discharge issuing from his member [*basar*], he is unclean."[27]

The book of the prophet Ezekiel (sixth century BCE) is especially well supplied with sexual metaphor, and there we find further evidence that *basar* sometimes means penis. In Chapter 16, the prophet condemns the Israelites for eagerly adopting alien customs: "You played the whore with the big penis [big *basar*] Egyptians," he declares. And in Chapter 23, in the course of another diatribe, again phrased in sexually explicit metaphor, Ezekiel castigates a woman symbolizing Jerusalem: "She lusted for concubinage with them [i.e., Babylonians], whose members [*b'saram*, plural of *basar*] were like those of asses and whose organs were like those of stallions."[28]

Of course, these biblical passages were all composed well before the rabbinic period. But there is good evidence that during the early centuries of Rabbinic Judaism the rabbis read some biblical references to *basar* as signifying not just flesh but penis. For example, the passage just quoted from Ezekiel 16 captured the imagination of an author cited in *Leviticus Rabbah* ("Great Leviticus," a book of rabbinic commentary based on Leviticus):

> What is meant by "great of flesh" [big *basar*]? Did this person have one leg and that three? The expression can only mean that they were all uncircumcised; and this is borne out by the text, "And the uncircumcised male who is not circumcised in the flesh of his foreskin" [Genesis 17:14]. Another exposition is that they possessed a *membrum virile* of abnormal size.[29]

So the rabbis were well aware that *basar* might mean "penis." And if the "life of the flesh" was indeed "in the blood," shedding even some of that blood constituted a sacrifice that could hardly fail to please Yahweh. I interpret this, then, as additional confirmatory evidence that the rabbis now envisioned circumcision as the new (and sole) blood sacrifice in their version of Judaism.

4. THE CONTEMPORARY RITE: REDEMPTION THROUGH BLOODSHED

Recent and contemporary Orthodox circumcision liturgies show that the ancient rabbinic connection between bloodshed and covenant has endured into our own time. The liturgies speak of the infant's genital blood as the contemporary equivalent of the blood displayed in temple sacrifices, portraying that blood as the substance that gains him entry into the covenant with Yahweh.[30] I'll cite three such liturgies: one from nineteenth-century England, and two from contemporary American sources.

In 1873, a prominent London physician named Asher Asher (sic) published a manual of Orthodox practice and interpretation entitled *The Jewish Rite of Circumcision*, presumably addressed to the growing Jewish population (both Sephardic and Ashkenazic) in England.[31] In a private prayer to be recited by the mohel (ritual circumciser) before the circumcision, the following appears: "Accept this my solemn service as erst Thou didst accept sacrifices on Thy altar"; and later he quotes the text from Zechariah mentioned earlier. Asher also records Sephardic prayers recited before and after the circumcision. The first includes this phrase: "O, have mercy on us and regard the blood of the Covenant." The second reads as follows:

> May the blood of the Covenant shed among you, O my faithful people, be acceptable to God as were the sweet smelling burnt offerings made in Jerusalem, His holy throne. He hath impressed His seal on our flesh for a sign and a token to us and to our children for ever; that all who see us, may know and recognize that we are the seed which the Lord hath blessed. May the blood of the Covenant shed among you, O my faithful

people, be acceptable to God as were the sweet smelling burnt offerings made in Jerusalem, His holy throne.[32]

Still widely used today is the text for the circumcision rite published in *Daily Prayer Book: Ha-Siddur Ha-Shalem*, by Philip Birnbaum, which Lawrence Hoffman describes as "a fixture in American Orthodox synagogues since its initial publication in 1949."[33] The mohel's principal prayer, recited immediately after he has circumcised the infant, includes the passage from Ezekiel 16 that is now standard for all Orthodox circumcisions: "I passed by you and saw you weltering in your blood. Live through your blood — I said to you — live through your blood."[34]

Moreover, the Birnbaum prayer book includes a reading to be added to the standard "Grace after Meals," recited in this case after the festive meal following the circumcision. The first verse reads as follows:

> May God bless this child's father and mother;
>
> May they bring him up and teach him wisdom.
>
> Henceforth may his blood win favor for him [literally, "From the eighth day and beyond may his blood be accepted].
>
> May the Lord his God ever be with him.[35]

Finally, there is *Bris Milah*, by Paysach J. Krohn, a standard text for ultra-Orthodox Jews, containing a lengthy rite with extensive commentary.[36] This includes a kabbalistic prayer to be recited by the mohel before the circumcision. One passage reads as follows:

> But now, because of our sins, our city [Jerusalem] has been destroyed, our Holy Temple consumed, and we have no sacrificial offering to atone on our behalf. May it be favorable [to You] to reckon the blood of this covenant as if I built an altar and brought onto it, burnt-offerings and sacrifices.[37]

The prayer in which the child receives his name also includes the reference from Ezekiel to living in (or through) one's blood, translated here (in bold type) as "and I said to you: 'Because of your blood you shall live!' and I said to you: 'Because of your blood you shall live!'" A note to the passage explains that the "double reference to blood refers to the two *mitzvos* [commandments], *milah* [circumcision] and the Pesach [Passover] offering, that God commanded the Jewish people to perform in order to merit redemption from Egypt. An uncircumcised person [i.e., man] may not eat from the Pesach offering and so it was necessary for all who had not been circumcised to perform this *mitzvah* [commandment]."[38]

In many communities, says Krohn, the mohel and father recite another prayer, which begins, "Master of the universe, may it be Your will that he be worthy, favored, and acceptable before You as if I had offered [literally, sacrificed] him before the throne of Your glory."[39]

The lengthy "Grace after Meals" in Krohn's text also includes the phrase, "from the eighth day onward may his blood be accepted." The explanatory note for the passage informs readers that some commentators explain "his blood" as referring to the father's blood shed at his own circumcision: "For, in reality, the child now being circumcised has not been asked if he is for or against being circumcised. ... Only when this same child will become a father and have his own son circumcised, will it become obvious that he is pleased that his parents had him circumcised. ... It is then that he is perceived as having performed the *mitzvah* of his own *bris* and we pray that *his blood be* [retroactively] *accepted*."[40] I find this unconvincing. It is certainly true that the hapless child has not been asked for his consent, and that the sacrifice is performed solely on the father's behalf, in the sense that it is his public performance (in the ancient past, perhaps saving him from having to sacrifice the child);[41] but the blood offered for Yahweh's acceptance is that flowing from the child's own genitals.

5. CONCLUSION: THE LIFE OF THE FLESH IS IN THE BLOOD

With the destruction of the Jerusalem temple and the end of Temple Judaism, the early rabbis realized that the only remaining candidate for a blood sacrifice was circumcision. The initial meaning of the circumcision rite — initiation into a covenantal community of males — they retained; but on this they superimposed a distinctly different meaning recalling ancient temple rituals: sacrifice of the male child's genital blood, to appease Yahweh and thereby redeem the boy from possible death. Thus, whereas in the initial (priestly) formulation of the rite, the definitive act was foreskin removal, the rabbis accorded primary recognition to what had previously been incidental but now became mandatory and purposeful: male genital bloodshed.[42]

Shedding that blood, and displaying it publicly, not only confirmed the covenant with Yahweh but also ensured that the unpredictable and sometimes malevolent deity would spare the child from the "pit of destruction." Although forced to sacrifice some blood from his own "flesh," the boy would not die a sacrificial death, but would live to become an adult member of a community of similarly consecrated and redeemed males.[43]

This may explain why *metsitsah*, sucking blood from the circumcised infant penis, was instituted as an essential third stage of the circumcision rite during the rabbinic period. Some have suggested that the blood is sucked to remove all vestiges of contaminating maternal blood from the child's body —credible enough in light of the phobic fear of female genital blood so characteristic of both Temple and Rabbinic Judaism.[44] Another possible explanation, though, is that sucking blood is likely to result in some appearing on the lips of the mohel — a vivid display of the core meaning of the rite.[45]

It may also explain why wine plays an unusual role in the circumcision ceremony. As Lawrence Hoffman has pointed out, ordinarily at Jewish ritual celebrations (Passover, Sabbath evening, and so on) one blesses wine, then drinks it. There is a cup of wine at a circumcision, but it is neither blessed nor drunk in the ordinary manner. Instead, immediately after the mohel has completed the recitation ending with the repeated phrase, "In your blood live," he dips a finger into the cup and places a drop or two of wine on the child's lips. "Now," explains Hoffman, "we see why no blessing anticipated this cup. It was not meant to be consumed *as wine* at all, but was instead reserved as an oral transfusion of wine *as blood* for the child. In a nutshell, blood escapes the system; wine as blood enters it."[46]

The rabbis who elevated circumcision to its new status as the sole blood sacrifice of post-temple Judaism knew that "the life of the *basar* is in the blood." Although unaware of blood circulation or the role of blood in erection, perhaps they reached a more fundamental conclusion: Did they realize that the penis *does* surrender some of its life when both its foreskin and a portion of its blood are removed? Was (and is) that the true purpose of Judaism's only surviving sacrificial rite?[47]

REFERENCES

[1]Letter to Toronto Globe and Mail, December 9, 2000. Dr. Jesin has a personal website, illustrated with a smiling infant (over the words "The Jesin Circumcision Clinic"), a physicians' caduceus, and "mazel tov!" in Hebrew script: http://drjesin.com/indexaj.html (March 4, 2001).

[2]The Talmud (the foremost text of rabbinic commentary, completed around 500 CE) includes a discussion of whether the "blood of the covenant" must be made to flow from infants born without foreskins and converts who were circumcised earlier: B. T. Yeb. 71a. Again, in Genesis Rabbah (a book of commentary on Genesis, composed at about the same time), several rabbis are cited in a discussion of the same question: Gen. R. 46.12, in Freedman H, Simon M, editors and translators. Midrash Rabbah: Genesis. London: Soncino Press; 1939. pp. 396-7. The essential references on circumcision and bloodshed are Hoffman LA. Covenant of Blood: Circumcision and Gender in Rabbinic Judaism. Chicago and London: University of Chicago Press; 1996; and Cohen SJD. A brief history of Jewish circumcision blood. In Mark EW, editor. The Covenant of Circumcision: New Perspectives on an Ancient Jewish Rite. Hanover and London: Brandeis University

Press; 2003. pp. 30-42. Cohen concludes that circumcision is now conceived to be blood sacrifice. Why, he asks, did circumcision blood become significant when and where it did? And why was circumcision not interpreted as sacrifice until then? This article may be read as an attempt to answer these questions, and to build on the pioneering research of both authors.

[3]For more detail on these and other biblical texts discussed in this article, see Glick LB. Marked in Your Flesh: Circumcision from Ancient Judea to Modern America. New York and Oxford: Oxford University Press; 2005. pp. 13-27.

[4]An instructive and readable reference on this complex topic is Friedman RE. Who Wrote the Bible? New York: Summit Books; 1987.

[5]Ex. 24:3-8; JPS Hebrew-English Tanakh. Philadelphia: Jewish Publication Society; 1999 (hereafter JPS). pp. 164-5; bracketed phrase is the literal meaning, cited in a note to the translation.

[6]Rogerson JW. Sacrifice in the Old Testament: problems of method and approach. In: Bourdillon MFC, Fortes M, editors. Sacrifice. London: Academic Press; 1980. pp. 45-59. [here, p. 59.] E. W. Nicholson also argues that the blood rite served not to ratify the covenant but for "consecration and purification." Nicholson EW. Exodus and Sinai in History and Tradition. Oxford: Basil Blackwell; 1973. pp. 72-4. As I've just suggested, these attributes might be better interpreted as complementary.

[7]JPS, p. 1395. This text is sometimes cited in prayers accompanying the circumcision rite.

[8]JPS, p. 211; "sprinkle of the blood" sic.

[9]Lev. 17:10-11; JPS, p. 248. An essential feature of *shechitah*, ritual slaughter to produce kosher meat, is that the animal must be exsanguinated. Orthodox Jews avoid blood consumption so strictly that even kosher meat must be salted and soaked to remove all traces of blood. In addition, an animal that proves on examination to have been infected or otherwise defective is unacceptable — just as was true for temple sacrifices. The Bible says nothing specific about ritual slaughter, aside from a single ambiguous passage in Deuteronomy: "you may slaughter any of the cattle or sheep that the Lord gives you, as I have instructed you ... But make sure that you do not partake of the blood, for the blood is the life, and you must not consume the life with the flesh" (Deut. 12:21, 23; JPS, p. 403). Julian Morgenstern notes that "early Semites," hunters and pastoralists, believed that the number of animals in each species was "definitely limited." To have consumed the blood would have meant consuming the soul, hence reducing the number of individuals in the species. This, he says, explains the biblical prohibition on blood consumption: "For the soul and the life were one; the soul was in the blood." Morgenstern J. The bones of the paschal lamb. J Amer Oriental Soc 1916;36:146-53. [here, p. 151.]

[10]Readers who feel that this statement requires proof might begin with Finkelstein I, Silberman NA. The Bible Unearthed. New York: Free Press; 2001. pp. 27-47.

[11]On circumcision among Semitic peoples as protection from evil spirits or redemption from child sacrifice, see Morgenstern J. Rites of Birth, Marriage, Death, and Kindred Occasions among the Semites. Cincinnati: Hebrew Union College and Chicago: Quadrangle Books; 1966. Chapter ix. For additional references on child sacrifice, see note 41 below.

[12]I discuss the relation between Genesis 17 and Genesis 15 in Glick LB. Marked in Your Flesh: Circumcision from Ancient Judea to Modern America. New York and Oxford: Oxford University Press; 2005. pp. 16-17.

[13]Blood does play a role in a cryptic circumcision narrative in Exodus. Moses's wife Zipporah circumcises their son to avert Yahweh's anger against Moses, touches the latter's genitals

also worth noting: "May God bless him who removed the foreskin,/And did fulfill all that has been ordained./One who is faint-hearted must not perform/This service, which includes three essentials." The "three essentials" are *milah*, severing and removing the foreskin; *peri'ah*, stripping away the mucosal lining and excising remaining tissue back to the corona; and *metsitsah*, sucking the bleeding penis, sometimes followed by spitting the blood into a cup of wine.

[36]I discuss this text in detail in Glick LB. Marked in Your Flesh: Circumcision from Ancient Judea to Modern America. New York and Oxford: Oxford University Press; 2005. pp. 218-20.

[37]Krohn PJ, editor. Bris Milah: Circumcision — The Covenant of Abraham. Brooklyn, NY: Mesorah Publications; 1985. p. 117; brackets and punctuation sic. Hebrew text on p. 116.

[38]Krohn PJ, editor. *Bris Milah*: Circumcision — The Covenant of Abraham. Brooklyn, NY: Mesorah Publications; 1985. pp. 131-32; my bracketed additions.

[39]Krohn PJ, editor. Bris Milah: Circumcision — The Covenant of Abraham. Brooklyn, NY: Mesorah Publications; 1985, p. 137; my bracketed translation of *hiqravtihu*. Cf. Hoffman LA. Covenant of Blood: Circumcision and Gender in Rabbinic Judaism. Chicago: University of Chicago Press; 1996. p. 72. In a note to this passage, Hoffman says, "The motif of sacrifice is itself a significant public meaning that I have chosen not to emphasize, because even though it does go back to rabbinic sources it becomes central only with the *Zohar*." He cites two references: Vermes G. Circumcision and Exodus IV 24-26. In: Vermes G. Scripture and Tradition in Judaism. Leiden: E. J. Brill; 1973, pp. 178-92; and Wolfson ER, Circumcision and the divine name. Jewish Quarterly Rev 1987;78:77-112. Wolfson notes that the theme of sacrifice "is repeated frequently in the *Zohar* [the foremost kabbalistic text]." [here, p. 99]. But although the theme may have received increased attention from thirteenth-century Jewish mystics, I think it is significant that it already appears in much earlier "rabbinic sources."

[40]Krohn PJ, editor. Bris Milah: Circumcision — The Covenant of Abraham. Brooklyn, NY; Mesorah Publications; 1985. pp. 149, 151; his punctuation, italics, and brackets.

[41]On the possibility (or probability) of child sacrifice in ancient Israel, see Levenson JD. The Death and Resurrection of the Beloved Son: The Transformation of Child Sacrifice in Judaism and Christianity. New Haven, CT and London: Yale University Press; 1993. esp. pp. 48-52 on circumcision; Day J. Molech: A God of Human Sacrifice in the Old Testament. Cambridge and New York: Cambridge University Press; 1989; and Glick LB. Marked in Your Flesh: Circumcision from Ancient Judea to Modern America. New York and Oxford: Oxford University Press; 2005. pp. 22-4. Julian Morgenstern says that, despite much scholarly opinion to the contrary, "the conclusion cannot be avoided that in Israel at one time the sacrifice of first-born children was a quite common, if not the regular, practice." Since the deity owned children and might take all of them if angered, it made sense to offer the first-born son as a sacrifice to redeem the others. He concludes that "first-born children were, beyond all question, regularly sacrificed to the proper deity or spirit as the natural and proper taboo-sacrifice": Morgenstern J. Rites of Birth, Marriage, Death, and Kindred Occasions among the Semites. Cincinnati: Hebrew Union College Press and Chicago: Quadrangle Books; 1966. pp. 63-4.

[42]It is obvious, of course, that our subject has been not just genital blood but *male* genital blood. I have room here only to mention the abhorrence of female genital blood so characteristic of rabbinic thought. Whereas male genital blood is believed to consecrate and to effect salvation, female genital blood — menstrual *and* that shed at childbirth — is viewed as a pollutant. (Nansi S. Glick has observed that female blood is acceptable

only in the display of bridal blood following a wedding night — also a traditional feature of many other Middle Eastern cultures, and yet another demonstration of male dominance.) See Hoffman LA. *Covenant of Blood: Circumcision and Gender in Rabbinic Judaism.* Chicago and London: University of Chicago Press; 1996; Archer LJ. 'In thy blood live': gender and ritual in the Judaeo-Christian tradition. In: Joseph A, editor. *Through the Devil's Gateway: Women, Religion, and Taboo.* London: SPCK; 1990. pp. 22-49; and Archer LJ. *Bound by blood: circumcision and menstrual taboo in post-exilic Judaism.* In: Soskice JM, editor. *After Eve.* London: Collins Marshal Pickering; 1990. pp. 38-61.

[43]*The Curse of Cain*, Regina M. Schwartz's perceptive study of the role of violence and ethnocentrism in the Hebrew Scriptures, begins with a chapter on covenants but (inexplicably) mentions circumcision only in passing. In one section of the chapter, called "Cutting Covenants," Schwartz refers to "the cutting of human flesh at circumcision — the so-called sign of the covenant." But, although she writes at some length about the cutting in Genesis 15, the original version of the myth of Abraham's covenant (in which sacrificial animals are cut to seal the covenant), she says nothing about Genesis 17, the revised version that incorporates the circumcision myth (where it becomes foreskins that must be cut). Moreover, another section of the covenants chapter, entitled "The Blood of the Covenant," discusses the blood-splashing episode at Sinai but still says nothing about circumcision. Schwartz RM. The Curse of Cain: The Violent Legacy of Monotheism. Chicago: University of Chicago Press; 1997. pp. 21-32.

[44]Note the talmudic reference to blood that is "stored up," and, in a passage from Maimonides, coming "from the distant places," quoted in Shields YP. The making of metzitza — 1972. Tradition 1972;13:36-48. [here, pp. 37-8.] Howard Eilberg-Schwartz remarks that "circumcision cannot be interpreted in isolation from the symbolism of blood generally and female blood particularly. Circumcision is a postpartum ritual associated with the separation of a male child from his mother. When the child is removed from the impurity of his mother's blood, he is brought into the covenant by the spilling of male blood. His blood is clean, unifying and symbolic of God's covenant. His mother's is filthy, socially disruptive and contaminating." Circumcision, he continues, thus "marks the passage from the impurity of being born of women to the purity of life in a community of men." Eilberg-Schwartz H. Why not the earlobe? Moment, February 1992;28-33. [here, pp. 32-3.] (These are the author's interpretations of religious ideology, not statements of personal belief.)

[45]Lawrence Hoffman discusses medieval texts describing men and adolescent boys dipping their hands and washing their faces and mouths with water into which has been dripped the blood of a newly circumcised infant: Hoffman LA. Covenant of Blood: Circumcision and Gender in Rabbinic Judaism. Chicago and London: University of Chicago Press; 1996. pp. 104-5.

[46]Hoffman LA. Covenant of Blood: Circumcision and Gender in Rabbinic Judaism. Chicago and London: University of Chicago Press; 1996. p. 91. At Passover seders, as the assembled group recites the ten plagues, it is traditional practice to dip a finger into a cup of wine for each plague and drip it onto a plate. The recitation begins with the first plague: *dam*, "blood," and concludes with *makat b'khorot*, "slaughter of the first-born [boys]."

[47]On this question, the great twelfth-century physician-philosopher Maimonides observed that, in his opinion, circumcision is performed "to bring about a decrease in sexual intercourse and a weakening of the organ in question, so that this activity be diminished and the organ be in as quiet a state as possible. ... For if at birth this member has been

made to bleed and has had its covering taken away from it, it must indubitably be weakened": Maimonides M. The Guide of the Perplexed. Pines S, translator and editor. Chicago: University of Chicago Press; 1963. p. 609. I discuss the complete passage in detail in Glick LB. Marked in Your Flesh: Circumcision from Ancient Judea to Modern America. New York and Oxford: Oxford University Press; 2005. pp. 64-6.

Chapter 3

ZIPPORAH AND THE BRIDEGROOM OF BLOOD

Searching for the Antecedents of Jewish Circumcision

Nansi S. Glick

INTRODUCTION

Anyone interested in the history of male circumcision in the ancient Near East, and particularly in its fateful adoption by the early Israelites, must be mystified by the available evidence in the Hebrew Bible. To be sure, the initial instructions for circumcision come early, in the seventeenth chapter of Genesis: to seal his covenant with Abraham, God requires that Abraham circumcise himself immediately along with every male of his household, slave or free, and ensure that his descendants henceforth circumcise all newborn males at the age of eight days.[1] Coming as it does so early in the biblical account of Israelite history, this passage seems to provide the basic template for Jewish circumcision ever since Abraham's putative lifetime, about 1,800 BCE. Yet there is surprisingly little evidence that Israelites in the distant past — until we reach a genuinely historical era, around 500 BCE — actually did practice male infant circumcision in the Abrahamic manner. Confirming Abraham's covenant we find merely a single sentence in Leviticus, in the midst of a series of regulations regarding female genital impurity after childbirth, mandating that a male child be circumcised at the age of eight days.[2] Beyond that, there is no further mention of *infant* circumcision in the Hebrew Bible.

There are, indeed, other biblical references to circumcision of boys or men, but they, too, are surprisingly few; and every instance, in its way, is puzzling. Dinah's brothers, by claiming that they are all circumcised, persuade the Shechemites to become so too; then the Hebrews are able to

G.C. Denniston et al. (eds.), Bodily Integrity and the Politics of Circumcision, 37—47.

slaughter them all in their weakened, post-operative state.[3] Ought we to believe that the Hebrew tricksters in this ancient tale were, in fact, all circumcised, and had been so since infancy? If so, what do we make of Joshua's command, in an equally ancient story, that all Israelite men undergo a "second circumcision" before their entry into the Promised Land? In this case, we're told that the second circumcision was necessary because the men "had not been circumcised" — which is odd if they had indeed been party to Abraham's covenant.[4] More straightforward was King Saul's challenge to David: bring home one hundred Philistine foreskins, and you can marry my daughter Michal. Since everyone knew that the enemy Philistines did not practice circumcision, this was meant as a daunting trial of strength; foreskins were specified as a customary — and pointedly macho — proof of success.[5] But none of these biblical mentions of circumcision tells us very much beyond the fact that the practice was known in pre-Judaic times. What evidence is there then, in the Hebrew Bible, to suggest the actual history of Israelite circumcision?

One enigmatic story remains to be mentioned: the account of the circumcision of Moses' son, sometimes referred to as the tale of the bloody bridegroom. This brief text in Exodus has been called "admittedly the most difficult of all the passages on this subject."[6] Commentators have sought to explain it from various points of view — theological, historical, psychoanalytic, ethnographic. But no one has analyzed the role of circumcision in the account; instead, it is always taken for granted as a bedrock Israelite custom for all male infants. Recent scholarship, however, allows us to consider a new possibility: that, at the time this text was written, universal male infant circumcision had not yet been mandated to the Hebrews and, therefore, that male circumcision had not yet become an ethnic marker for them. In light of this, we can begin to explain this puzzling tale — and, by doing so, illuminate the likely role of circumcision in the pre-Judaic world.

1. DATING THE TEXT

First we need to look at the tantalizingly brief text itself. Moses, whose birth and upbringing in Egypt are recounted in the first three chapters of Exodus, killed a man and fled to Midian, where he married Zipporah, daughter of a Midianite priest. Now they have two sons. One day, Yahweh appears to him in a burning bush and commands him to return to Egypt to confront a new pharaoh and lead the Hebrew people to freedom. After briefly protesting his unsuitability for the task, Moses sets his wife and young sons on a donkey and begins the journey to Egypt. At this point, our story suddenly appears, without introduction or apparent context:

> At a night encampment on the way, the Lord encountered him and sought to kill him. So Zipporah took a flint and cut off her son's foreskin, and touched his legs with it, saying, "You are truly a bridegroom of blood to me!" And when He let him alone, she added, "A bridegroom of blood because of the circumcision."[7]

The episode ends there; the main narrative proceeds as though nothing remarkable has happened.

Part of the difficulty of this account lies in the Hebrew text itself, which (as the translation demonstrates) does not specify which of the three male actors — Yahweh, Moses, or the boy — is indicated by the masculine pronouns. This has led interpreters to differ on the essential matter of who does what to whom; and analysts have supported each of the gamut of possibilities.[8] Thus even the action of the story is unclear. For reasons that will be explained, I think the only way to make sense of the tale is to assume that events occurred as follows: Yahweh is angry at Moses and seeks to kill *him*.[9] Zipporah, to save her husband's life, circumcises their son and touches Moses with the bloody foreskin; I'll deal later with precisely where she touches him. Their son, a passive actor in the story, merely provides the foreskin.

Now, since this passage appears in the book of Exodus, anyone reading the Bible linearly must assume that Moses, as an ostensible heir to the Abrahamic tradition originating several centuries earlier, must have known that his sons ought to be circumcised. Most readers have taken this for granted; indeed, until quite recently, they could hardly do otherwise. The resulting confusion compounds the difficulty of deducing the motives of the actors. If Yahweh is attacking the boy for being uncircumcised, why does he choose this occasion to express it — and why doesn't Moses do the circumcising? If he's angry at both parents for forgetting their ritual obligations, why does Zipporah not circumcise both sons? For that matter, why is a woman, and a Midianite woman at that, doing the circumcising? Should we assume that Zipporah touches Yahweh with the foreskin, or even the boy? Or does she smear Moses with it to acknowledge that he himself remains erroneously uncircumcised? And what does she mean by repeating the phrase, "bridegroom of blood?" Is she addressing the boy? If so, might circumcision have been a prerequisite for marriage in those days? But then, if this phrase had indeed been part of a customary incantation for the circumcision ritual, why must she circumcise her young son tonight, just as his father is about to undertake the most momentous actions of his life? All these questions have been argued, but no one has presented a convincing, unified explanation for both parts of the story: the emergency circumcision and the "bridegroom of blood" statement.

On only one point every analyst agrees: this mysterious tale is very old. No one, however, has compared its age — *as a text* — to that of Chapter 17 of Genesis, in which Abraham receives the Lord's command to circumcise himself and his descendants. When we make such a comparison, we find that we are dealing with a kind of circumcision very different from the Abrahamic model.

Modern scholarship has shown that scrolls containing mythical and religious writings of the Hebrew people were created at various times starting around 900 BCE.[10] At least three major texts, composed at intervals of a century or two, were circulating in the Israelite community by the time the Judeans returned from their Babylonian exile, shortly before 500 BCE. Their priestly leaders, determined to create an all-encompassing religious polity dedicated to the exclusive service of the single god Yahweh, set about collecting and organizing all the extant writings relating to the Hebrew people's origin, history, and relationship with Yahweh. An editor or editors collated origin myths and tales of very early times into a book that became Genesis; accounts of the departure from Egypt became Exodus; and so on. But the editors, as members of a newly-constituted hereditary priesthood, were not content simply to collect existing texts: they added to them, amended them, and, in many cases (especially in Leviticus, Numbers, and Deuteronomy) included entirely new material of their own. Scholars label these priestly additions the "P" text. The assembled writings — ancient myths mingled with priestly regulations — became the foundational five biblical books known as the Torah.

One result of recent textual studies is that almost all passages of the Torah can now be approximately dated. Passages that appear side by side in a single chapter may, we now know, have been written several centuries apart. Thus we learn that the bridegroom of blood tale is among the earliest, dating to sometime around 900 BCE. But the account of Abraham's circumcision —*even though it appears in the first book of the Bible and purports to tell of a much earlier time in Hebrew history* —was actually written by priests about 500 BCE.[11] To be sure, the priests based their account of Abraham's covenant on an existing narrative; and, in fact, a version of the Abrahamic covenant story written around the time of the Zipporah tale, though with no mention of circumcision, is still to be found in Genesis 15. But it bears repeating that the universal circumcision mandate — that all Israelite males must be circumcised at the age of eight days — was added to the Abraham narrative several centuries after the tale of the bridegroom of blood first appeared.

Thus it follows that the original author of the bridegroom of blood text, along with his contemporary readers, knew nothing of mandatory infant circumcision for Hebrew boys. In the context of his own time, he assumed that his audience would understand what Zipporah meant to do when she circumcised her son that fateful night. Our challenge is to recreate what they took for granted.

2. CIRCUMCISION AS SACRIFICE

When Jacob wrestled with the Lord, he was alone, with no one to call on for assistance; he emerged from the night-long ordeal alive but wounded.[12] But when Yahweh assaulted Moses, Zipporah was on hand to help.[13] The text makes clear that circumcising their son was her attempt to mitigate Moses' peril. Since ritual circumcision of infant boys was unknown at the time, her actions could not have been spurred by any guilty realization that her son ought to have been circumcised already. Why, then, in response to her husband's danger, did she decide to perform an emergency circumcision? The answer can only be that she understood circumcision to be a *sacrifice*. And to offer a meaningful sacrifice was the only way she knew to appease an angry deity.

Sacrifice was thoroughly familiar to the ancient Canaanites, including the early Israelites. It was, in fact, their primary way of dealing with capricious and often irascible deities. There was no prayer, no code of personal conduct that might serve to assuage a deity's anger; all that came later in the history of Israelite religion. Sacrifice, minor and major, was the foundation of relationships with deities throughout the region.

The most common sacrifice was a domestic animal, usually a sheep or goat; for lesser occasions a bird, fine grain meal, or oil might suffice. This was also a society that countenanced child sacrifice; in the same way that sacrifice of first-born animals ensured future fertility of the dam, sacrifice of first-born children (perhaps only sons) likewise ensured a large family thereafter.[14] But, as one might expect, there soon arose mechanisms to "redeem" a child from sacrifice.[15] Furthermore, on the old principle of a part for the whole, it became possible to sacrifice a significant part of the child rather than the child himself. What, then, was the most significant part of a child? In a patriarchal society, where sons were a man's most valued possession — for their potential to produce sons of their own and thus perpetuate the line — the boy's genitals embodied all the dearest values of patriarchy. To cut off part of a boy's penis was thus — from his *father*'s point of view — a major sacrifice; it came close to sacrificing the child himself, reminding the deity that this represented the father's most precious asset. The sacrifice, then, was never performed on the boy's behalf, any more than a bull was sacrificed for its own sake. Men who commanded both herds and households could choose a sacrifice to suit their needs and the disposition of the deity who had to be placated.

Anthropologists suggest that sacrifice is particularly common among pastoral, patriarchal peoples, and it is still practiced in much of the Middle East and North Africa today.[16] They point out that a sacrifice is a kind of gift, and gifts always assume a return. In other words, to offer something of value to a deity *obliges* the god to respond with supernatural assistance to the person who makes the sacrifice. The good behavior or moral worth of the sacrificer has no bearing on the power of the sacrifice: *the deity cannot refuse*.[17] (Of course, there were occasions when a sacrifice seemed to have been made in vain; but in that case it was assumed that the deity, not the sacrifice, was at fault, and supplicants could switch their allegiance to a more effective god.)

Thus Zipporah is confident that Yahweh, whatever the cause of his anger, will respond favorably to her sacrifice of Moses' son. Picking up the household flint knife, she performs a simple amputation of the tip of her son's foreskin; then she turns away from the boy. His role has ended; he has been circumcised for his father's sake. But the sacrifice is not yet complete: Zipporah must now display the sacrificial blood where Yahweh will see it.

Blood was and is the essence of sacrifice. The ancient Israelites slaughtered animals for food by exsanguination — slitting the throat so that the animal bled to death — and this practice is still followed by Jewish kosher slaughterers. A sacrificial animal was killed in precisely the same way; *it became a sacrifice only when its blood was appropriately displayed before the deity*. Intention alone does not make a sacrifice. What matters is ritual disposition of the blood; only then will the deity respond.[18]

Biblical regulations specified that priests who sacrificed on their own or their congregants' behalf to atone for a personal transgression, were to smear some of the blood on their own persons — on their earlobes, toes, clothing — as well as on the altar and other more traditional locations.[19] Similarly, foreskin removal became a sacrifice only when it was offered as such — when it included display of blood before a deity. Without that, circumcision was a gesture (and a forceful one) of domination, as the tales of David's killing of Philistines and the massacre of the Shechemites indicate.[20] Zipporah surely understood that the final, confirming step of her sacrifice must be to display the blood — and to do so on the body of the transgressor, Moses. Without needing to know *why* Yahweh was so murderously angry, she could be certain that her sacrifice would oblige him to favor rather than harm her husband.

This is why she used the severed foreskin to smear blood on Moses himself. Now, where on his body should she choose to make the display? The translation says that she touched his "legs" with the bloody flesh. Although we might take this statement literally — the Hebrew word is *raglayim*, meaning legs or lower limbs — the term was often used in biblical contexts as a euphemism for penis. Here it makes sense to understand the word that way.[21] We must imagine Zipporah smearing Moses' penis with the gory foreskin, painting it with blood. At once, without explanation or ceremony, the Lord withdraws, leaving Moses unharmed. Her sacrifice has succeeded.

Now Zipporah delivers the puzzling observation that Moses has become "truly a bridegroom of blood" to her. If my analysis of the circumcision sacrifice is correct, this is readily explained. There could have been only one other occasion when she had seen him with a blood-streaked penis: on their wedding night. It was supremely important then, as it still is today in many Mediterranean societies, that a woman be a virgin at her wedding. (In fact, for a husband to deflower a virgin bride, ensuring a pure line of descendants, was itself a kind of sacrifice.[22]) Thus a proud new husband, displaying the proof of his blood-smeared penis, might boast of being a bridegroom of blood.[23] And now the blood on Moses' penis reminds Zipporah of this earlier time: "A bridegroom of blood because of the circumcision."

3. CONCLUSION

What can we learn about the earliest practice of circumcision from this ancient tale? We know now that Jewish circumcision, the rite that initiates all eight-day-old males of Jewish descent, began *late* in Hebrew history, around 500 BCE. It was not until the return from Babylonian exile that organized Judaism, as we recognize it from the Hebrew Bible, began. By the time the Persian conquerors of Babylon allowed some exiles to return to Jerusalem, Judean scholar-priests had already worked out the complex codes of religious conduct that they had determined were essential for the new Judean community. They saw the return to Jerusalem as a unique opportunity to implement new rules for a rigorous theocracy, and to control, for the first time ever, the private lives of their entire (though admittedly tiny) society. It was then that they mandated infant circumcision for all Judean males.

But before that time, circumcision played an entirely different role. The practice was widely known; in fact, for a long time it was considered primarily an Egyptian custom, evidently an initiation ritual for priests. But Canaanites, strongly influenced by Egypt, practiced circumcision too. As far as we can tell, it was never universally mandated by any group. Egyptian documents mention circumcised enemy combatants, but they are never associated with any particular ethnicity. *Some* Canaanite men, including Hebrews, seem to have been circumcised, but never *all* males of any particular people.

It appears, therefore, that in ancient Canaan, before 500 BCE, circumcision must have been considered a sacrifice. It would have been performed on behalf of a householder — a father — who owned wives, offspring, and flocks, and who might on occasion need the powerful supernatural assistance that a circumcision sacrifice provided. So some boys in an extended household might be circumcised, and others not; perhaps first-borns were more likely to have been chosen. A circumcised man, whose penis had been thus ritually sacrificed, might have been considered sanctified to the deity (and perhaps, therefore, eligible to become a priest). The "second circumcision" of the Israelites just before they crossed into the Promised Land might then be explained as Joshua's way of sanctifying the men who were to enter Yahweh's domain.[24]

Since priests had not been a hereditary caste before the return from exile, priestly legitimacy might have been under challenge. Thus a mandate to circumcise *all* males, sanctifying the entire Jewish polity to Yahweh, would leave no one able to claim special circumcised status. And requiring that *infants* be circumcised ensured that no Jewish father could later perform his own personal circumcision sacrifice. All other sacrifices could then be consolidated in the Temple under priestly control. Meanwhile, the perilous circumcision of infant boys obliged their fathers to acknowledge the primacy of the Jewish community into which their sons were being initiated.

It was the priests, then, with their mandate to circumcise boys at the age of eight days, who profoundly changed the meaning of circumcision. Once *every* male infant had to be circumcised, individual sacrifice obviously lost its significance. And after only a generation or two, Judean males came to be seen as "marked" by circumcision. Though occasional circumcision of boys, as sacrifice, may have continued for a time in other Canaanite cultures, universal *infant* circumcision soon became the most widely recognized Jewish practice — and the distinguishing mark of all Jewish males.

REFERENCES

[1]Genesis 17:10-14. JPS Hebrew-English Tanakh. Philadelphia: Jewish Publication Society; 1999. p. 29.

[2]Leviticus 12:3. JPS Hebrew-English Tanakh. Philadelphia: Jewish Publication Society; 1999. p. 231.

[3]Genesis 34. JPS Hebrew-English Tanakh. Philadelphia: Jewish Publication Society; 1999. pp. 70-2. The text specifically notes that Jacob's sons spoke "with guile," v. 13, p. 71, suggesting that their claim that they were all circumcised was misleading. This text dates from about 900 BCE. [Friedman RE. Who Wrote the Bible? New York: Summit Books; 1987. p. 248.]

[4]Joshua 5:2-8. JPS Hebrew-English Tanakh. Philadelphia: Jewish Publication Society; 1999. pp. 464-5. This passage seems to have been written sometime around 700 BCE. [Finkelstein I and NA Silberman. The Bible Unearthed: Archaeology's New Vision of Ancient Israel and the Origin of Its Sacred Texts. New York: Free Press; 2001. chap. 3.]

[5]I Samuel 18:25-27. JPS Hebrew-English Tanakh. Philadelphia: Jewish Publication Society; 1999. p. 614. David returned in triumph with 200 foreskins, though in II Samuel 3:14, p. 649, he claims he paid only "one hundred Philistine foreskins" for Michal.

[6]Isaac E. The enigma of circumcision. Commentary 1967;43(Jan.):51-5. [here, p. 53.] He goes on to say that "[t]he interpretations of this passage have usually been even more peculiar than the passage itself."

[7]Exodus 4:24-26. JPS Hebrew-English Tanakh. Philadelphia: Jewish Publication Society; 1999. p. 120.

[8]Most authors provide a critique of their predecessors before attempting to explain the tale themselves. See especially Isaac E. The enigma of circumcision. Commentary 1967:43 (Jan.):51-5; and Eilberg-Schwartz H. God's Phallus and Other Problems for Men and Monotheism. Boston: Beacon Press; 1994. pp. 151-61.

[9]For the most part, I follow the lead of Howard Eilberg-Schwartz, whose book, God's Phallus and Other Problems for Men and Monotheism, contains the most perceptive analysis of the Zipporah story. Eilberg-Schwartz and others maintain that Yahweh attacked Moses with the intention of damaging his genitals, just as he (or his angel) wrestled with Jacob — wounding him in the groin — when he, too, was on his way to assume leadership of the Hebrew people. They say Yahweh insists in these episodes that Israelite leaders not be too cocky (to use a word they avoid) in their relationship with the Lord: that the Israelites should assume a feminine role in their receptivity toward God, though masculine enough towards the outside world. Though this is presented as a theological argument, it has the additional advantage of clarifying the action of the Zipporah story. Eilberg-Schwartz H. God's Phallus and Other Problems for Men and Monotheism. Boston: Beacon Press; 1994. pp. 151-61. See especially p. 161, where he states that "circumcision was for the ancient Israelites a symbol of male submission [to God]."

[10]For recent works in this field, see Friedman RE. Who Wrote the Bible? New York: Summit Books; 1987 and Finkelstein I and NA Silberman. The Bible Unearthed: Archaeology's New Vision of Ancient Israel and the Origin of its Sacred Texts. New York: Free Press; 2001.

[11]Friedman RE. Who Wrote the Bible? New York: Summit Books; 1987, gives the earliest date (circa 900 BCE) for the Zipporah story (p. 250). Friedman assigns the Genesis 17 command to circumcise all newborn boys — a P text — to around 500 BCE (p. 247). For further discussion of this dating of the circumcision mandate, see Glick LB. Marked in Your Flesh. New York and Oxford: Oxford University Press; 2005. pp. 14-7.

[12]Genesis 32:25-32. JPS Hebrew-English Tanakh. Philadelphia: Jewish Publication Society; 1999. p. 68.

[13]One article on this story, in fact, is titled "Zipporah to the Rescue." (Robinson BP. Zipporah to the rescue: a contextual study of Exodus IV 24-6. Vetus Testamentum 1986;36(4):447-

61.) Robinson recognizes her service to Moses, but cannot fully explain it. Not all commentators have understood the passage this way, because there is no literal basis in the Hebrew text for the assumption that Zipporah's actions came in response to her husband's peril. Although the translation says "*So* Zipporah took a flint," no motivation or causality can be inferred; the Hebrew word for that transition is the bland "and" that merely indicates sequential action.

[14]In addition to the near-sacrifice of Isaac (Genesis 22), there are many biblical injunctions similar to Exodus 22:28: "You shall give Me the first-born among your sons." JPS Hebrew-English Tanakh. Philadelphia: Jewish Publication Society; 1999. p. 161. For further discussion, see Day J. Molech: A God of Human Sacrifice in the Old Testament. Cambridge and New York: Cambridge University Press; 1989; Levenson JD. The Death and Resurrection of the Beloved Son: The Transformation of Child Sacrifice in Judaism and Christianity. New Haven: Yale University Press; 1993; and Glick LB. Marked in Your Flesh. New York and Oxford: Oxford University Press; 2005. pp. 22-4.

[15]Exodus 13: 13, 15. JPS Hebrew-English Tanakh. Philadelphia: Jewish Publication Society; 1999. p. 140.

[16]See, for example, Jay N. Throughout Your Generations Forever: Sacrifice, Religion, and Paternity. Chicago and London: University of Chicago Press; 1992. See especially her introduction, pp. xxiii-xxvii.

[17]The timely use of sacrifice to avert disaster is well illustrated in a biblical account, written about the same time as the Zipporah story, of a battle involving Mesha, king of Moab. His opponents, kings of Israel, Judah, and Edom, already had Yahweh's blessing through the prophet Elisha, and Yahweh's assistance had brought them inexorably close to total victory. Thereupon Mesha decided that he needed supernatural aid of his own. He "took his first-born son, who was to succeed him as king, and offered him up on the wall as a burnt offering. A great wrath came upon Israel, so they withdrew from him ..." Mesha's deity was Chemosh, though the Israelite authors omit mention of this. The point was not that Chemosh's power proved greater than Yahweh's, it was that Mesha's *sacrifice* was greater. And it handily excused the Israelites' defeat. II Kings 3: 26-7. JPS Hebrew-English Tanakh. Philadelphia: Jewish Publication Society; 1999. p. 781.

[18]Leviticus 17:11-14. JPS Hebrew-English Tanakh. Philadelphia: Jewish Publication Society; 1999. p. 248. See also Leviticus 19: 26, p. 252.

[19]Leviticus 14: 14, 25. JPS Hebrew-English Tanakh. Philadelphia: Jewish Publication Society; 1999. pp. 238, 239. These are instructions to cleanse a leper with a "guilt offering": "When the lamb of guilt offering has been slaughtered, the priest shall take some of the blood of the guilt offering and put it on the ridge of the right ear of the one being cleansed, on the thumb of his right hand, and on the big toe of his right foot." (Leviticus 14:25, p. 239). See also Leviticus 8:23-4 and 30, pp. 223 and 224.

[20]I Samuel 18: 25-7, Genesis 34. JPS Hebrew-English Tanakh. Philadelphia: Jewish Publication Society; 1999. pp. 614, 71. See also Eilberg-Schwartz's comments on domination, cited in note 9 (above).

[21]Eilberg-Schwartz, among others, argues for this translation for theological reasons; see note 9, above. Eilberg-Schwartz H. God's Phallus and Other Problems for Men and Monotheism. Boston: Beacon Press; 1994. pp. 151-61. If we follow his lead and assume that the word here means penis, we are able to explain Zipporah's references to the "bridegroom of blood."

[22]M.E. Combs-Schilling argues precisely this in a chapter, "Bride's Blood," of her book, Sacred Performances. (Combs-Schilling ME. Sacred Performances: Islam, Sexuality, and Sacrifice. New York: Columbia University Press; 1989.) She comments that "anthropologists

have been off the mark in single-mindedly concentrating on females and virginity in Mediterranean marriage practices, as opposed to males and their act of bloodspilling; it is striking how male-focused is the ritual ...virginity, like the death of the animal in sacrifice, is a secondary rather than a primary factor in these rites ...What is important is that the men spill blood in an act of some violence, so that they demonstrate for themselves and others their procreative role" (p. 209). She notes that, furthermore, "the groom must spill sacrificial blood, but *this can be accomplished through substitute sacrifice*" (p. 209; my emphasis); the substitute could be a bird smuggled into the bridal chamber and formally sacrificed over the bride's genitals. Of her observations in Morocco (though we have to remember that this is contemporary evidence, not ancient), she says, "virginity is *not* always demanded, whereas bloodspilling is, a clear indication of what is culturally primary and what is secondary" (p. 210; her emphasis). Thus Zipporah's circumcision of their son may have constituted a dual sacrifice. In the first place she has appeased Yahweh's anger. And, as her comments seem to indicate, by smearing the sacrificial blood of the circumcision on her husband's penis, she has, in effect, performed a *substitute sacrifice* that recreates Moses as "truly" a bridegroom of blood.

[23]Deuteronomy illustrates how the virginity of a bride could be proved by preserving the spilled hymenal blood on a nuptial cloth. If a married woman's parents must defend their daughter's virginity at marriage, they need only produce the linen, declaring "'...here is the evidence of my daughter's virginity!' And they shall spread out the cloth before the elders of the town." [Deuteronomy 22:16-17. JPS Hebrew-English Tanakh. Philadelphia: Jewish Publication Society; 1999. p. 422.] This passage was composed a couple of centuries after the Zipporah narrative. Before nuptial bedlinens became common, the proof of a bride's virginity might well have been her new husband's blood-smeared penis, displayed before selected witnesses.

[24]Joshua 5:2-8. JPS Hebrew-English Tanakh. Philadelphia: Jewish Publication Society; 1999. pp. 464-5.

Chapter 4

AT THE ROOTS OF ETHNIC FEMALE GENITAL MODIFICATION

Preliminary Report

Pia Grassivaro Gallo, Eleanora Tita, FrancoViviani
Padua Working Group on Female Genital Mutilation, Department of General Psychology, University of Padua (Italy)

Abstract: The substitution of the term female genital mutilation with ethnic female genital modification (EFGM) would be beneficial because of its lesser semantically negative connotation and superior expression of the interventions' expansive/reductive duality. All forms of EFGM have a clear relationship with coupling, and thus with the resulting fecundity (concrete manifestation of procreative capacity). Fecundity is the fundamental element linking the first inhabitants of Africa to their most recent descendants (De Rachewiltz, 1963). All forms of EFGM aim to increase fecundity (in reality or according to cultural assumption) and are accomplished by either enhancement of genitalia (expansive modification, such as elongation of the labia minora) or degradation of genitalia (reductive modification, such as infibulation). Both longininfism and infibulation refer, in their origin, to the same past population, the Bushmen/Hottentots. This group carried the "apron" as a congenital, morphologic feature. This has been preserved through manipulation, for example, among the Baganda of Uganda. In Somalia, the Midgan, operators of infibulation, belong to an ethnic group of hunters (the Bon), probably descending from an ancient Khoisanid population.

INTRODUCTION

Since 1996, the World Health Organization (WHO) has used the term female genital mutilation (FGM) to classify a group of permanent interventions on the female external genitalia proving extremely heterogeneous in nature, for their consequent repercussions, and in the socio-cultural systems of the populations to which they refer. Use of the term "mutilation" is

G.C. Denniston et al. (eds.), *Bodily Integrity and the Politics of Circumcision,* 49—55.

inappropriate in several cases belonging to type 4, and assumes a strongly negative semantic connotation especially for East-African excised immigrants, who do not consider themselves mutilated. The resulting discomfort of researchers and operators in regard to this generalized expression is increasingly evident in recent studies (Shell-Duncan and Hernlund, 2000; Gruenbaum, 2001; Fusaschi, 2003), where various alternatives are proposed.

1. REDEFINING FGM

It is our opinion that the most adapted alternative expression is ethnic female genital modification (EFGM), as seen in several recent (and less so) works (Puccioni, 1904; Erlich,1986; Fusaschi, 2003). The term "modification" is acceptable both to western women and African immigrants. Furthermore, use of the adjective "ethnic" underlines, in addition to the plurality of excisory populations, the cultural rather than therapeutic motivation of these interventions, as emphasized in the WHO definition.

The proposed expression can also refer to the expansive/reductive duality of these interventions (Fusaschi, 2003). This dichotomy of the EFGM stands confirmed by the geographic distribution of the interventions in Africa, which we have mapped in reference to longininfism (ritually induced hypertrophy of the labia minora). (Map), in respect to the well-noted distribution of reductive forms as defined by Hosken (1982; 1994); examining the linguistic and ecological elements. Expansive populations have at disposition a rich and articulated vocabulary for topics within the sexual sphere (Kashamura, 1973; De Rachewiltz, 1963; Grassivaro Gallo and Villa, 2004); reductive populations, on the other hand, have a minimalist vocabulary lacking many expressions and concepts tied to sexual life (Taha Baasher, 1982).

From an ecological standpoint, East African infibulators are primarily nomadic pastoralists (Hicks, 1986; Grassivaro Gallo and Viviani, 1992), and the expansors of the Great Lakes region are prevalently agricultural populations (Grassivaro Gallo and Villa, 2003).

2. CULTURAL ROOTS OF THE EFGM

All forms of EFGM are indispensable prerequisites to marriage. This clearly influences fertility, and thus its practical realization for every woman, fecundity, defined as the concrete manifestation of procreative capacity (Livi Bacci, 1990).

The aim of EFGM is to increase fecundity (the number of children). This, in fact, is achieved by expansive interventions (such as longininfism), and is believed by cultural assumption to do so also for reductive interventions, which sever the external genitals, and thus degrade the sexual act (as seen in infibulation).

Fecundity is the most significant link between the earliest inhabitants of Africa and their progeny, the current inhabitants of the dark continent (De Rachewiltz, 1963). EFGM serves as a clear example of heredity of cultural characteristics that are passed down through generations by laws equally rigid to those of biological heredity (Cavalli-Sforza, 2002).

Both expansive and reductive forms of EFGM refer to the same ancestral population, the Bushmen/Hottentots, the presence of whom in Africa dates back to the Paleolithic (Battaglia, 1954).

3. THE BUSHMEN/HOTTENTOTS (B/H) AND THEIR HEREDITY

These early inhabitants, described since the 1700s and 1800s by anthropologists as hunter/gatherers of Northeast Africa (Baumann and Westermann, 1948) no longer exist as a pure group. The representatives of the Boscopoid race, believed to have preceded them, are completely unknown to the anthropological record. The Bushmen/Hottentots, however, have been described in great detail, especially considering their morphological peculiarities and specific living conditions (Le Vaillant, 1790; Cuvier, 1817; Stow,1905; Peron and Lesueur, 1883; Schultze-Jena, 1928; Rugiu, 1931).

Between the physical traits is the structure of the male and female external genital apparatus, clearly conducive to the sexual act. The men maintained a penis rictus (of short length, normally in an erect position), while women presented a congenital hypertrophy of the labia minora (apron, grembiule, tablier), accompanied by reduced salience of the Venus mount and labia majora, thus allowing the enlarged labia to compensate for the reduced penis size, and facilitate coitus (De Rachewiltz, 1963).

Pushed by more modern groups, the Bushmen/Hottentots began to migrate towards the central and southern areas of the continent, always towards the East (Stow, 1905; Schapera, 1930; Seligman, 1930; Bleek and Lloyd, 1968). Coming into contact with resident populations, they hybrid-dized to the point of disappearance as a pure race. Currently, small residual nuclei dispersed throughout western Africa remain, included among the ethnicities most often subjected to manual labor (Kindiga e Sandawe, del Tanganika; Ndorobo, Midgan e Bon, della Somalia; Wata del Nord-Est;

Sanye-Ariangolou, in Africa Orientale; Gow del Niger; Mahalbi, del Niger e Chad; Nemadi del Giuf – Baumann and Westermann, 1948; Battaglia, 1954; Grottanelli, 1976).

Through hybridization, the tablier as a recessive trait (Peron and Lesueur, 1883; Dart, 1937) was lost (as well as the male penis rictus). The memory of its facilitation of procreativity, however, was maintained through cultural heredity. Contemporary populations of the Great Lakes region ritually obtain elongated labia minora, associated often with a corresponding enlargement of the clitoris and vaginal canal substantial enough to achieve the same result as the tablier, obtaining maximum satisfaction from the sexual act, thus increasing reproductive success.

The link between the Bushmen/Hottentot and expansive EFGM is thus incontestable.

Passage to the reductive forms of genital modification is less evident, and requires an introduction.

The labia minora are erectile structures (similar to the clitoris), characterized by ample dimensional variety between individuals. Occasional cases of hypertrophic labia minora have been noted also among modern European populations, which are considered pathologies in gynecological literature (Jeffcoate, 1967); when this occurs, modern aesthetic surgery contradicts the condition through labiaplasty (Guth and Von Samsonov, 2001) in cases where the patient complains of hypersensitivity due to the large labia minora (which may also result from excessive masturbation).

In the same vein, it is possible that peri-saharian populations began to systematically reduce the dimensions of the labia minora and clitoris, becoming excisors. Through oral tradition, a series of negative connotations regarding these features was then maintained.

In this evolution, one should also consider the fear regarding a woman so powerful as to manage the success of a sexual relationship (Fussaschi, 2003). It is certain that in Ethiopia (Bruce, 1791), and in Egypt (Sonnini, 1798) excision has been described as an intervention necessary to prevent the "abnormal" development of genitalia left.

In all of East-Africa, these are considered ugly and obscene; it is believed that non-excised, the labia and clitoris may grow to the point of dangling between the legs (as would a penis); furthermore, contact between the clitoris and the male organ during coitus, or a baby at birth, is perceived as potentially fatal. A non-excised woman may also prove to be sexually uncontrollable.

Recent studies, however, have not found proof in favor of the assumption of hypertrophy of the labia minora in North-East Africa (Erlich, 1986; Hosken, 1982).

Concluding, also the origin of excision can be connected to labial hypertrophy among the Bushmen/Hottentot, the only group among which it has been described as a congenital feature.

Infibulation, the focus of which is found in the Horn of Africa, from where the Bushmen/Hottentot began their migration, may have begun casually as an accidental side-effect of excisory practices (a not-rare occurrence even today (Grassivaro Gallo, 1986: interview F, page 186). Secondarily, it may have been adopted systematically due to its usefulness in a pastoral environment to protect women, children, and herds of sheep from predators (Grassivaro Gallo and Viviani, 1992).

It is additionally necessary to consider that, at least in Somalia, traditional operators of infibulation, the Midgan, belong to a group of hunters from Low Giuba, the Bon, descendant from the Bushmen. They perform these interventions on the daughters of their bosses, but leave their own intact. It is possible that they, profoundly knowledgeable on nature, herbs, and the medical arts, are the inventors of the most severe genital mutilation.

In every attempt to analyze genital modifications, the argument always returns to the Bushmen/Hottentot.

4. CONCLUSIONS

From the analysis executed, one may note foremost the depth of the roots of EFGM in time, which have spread through generations as a cultural characteristic since the Paleolithic.

Fundamentally correlated to these practices is access to marriage, anticipated by all for excised women. The centrality of these practices has been clearly elaborated in recent studies following socio-political directives extraneous to our analysis (Mackie, 2000), arriving at the same conclusion. Marriage is viewed as the institution in which one develops legitimate progeny that must be as ample as possible.

Fertility in general and the fecundity of every woman thus serves as the motor of the EFGM, in both its reductive and expansive forms. Considering these indications, we may also gain understanding of the tenacious affection held by the African woman for these practices, otherwise inexplicable to the western observer. She feels ancestrally obliged to sustain this tradition to prevent the extinction of her progeny and the disappearance of her ethnicity. On this point, the substantial incomprehension of western society, completely estranged to these values, and among western women, the maximum incomprehension may be attributed to the Italian woman, who has the lowest birth index of all.

Precursors of these traditions, the Bushmen/Hottentot migratory directives in Africa perfectly coincide to the eastern territories where modern day genital modifications are performed, thus demonstrating them to be a case of convergent cultural evolution; while the reductive/expansive duality of the modality in which they are realized appears to be a case of divergent cultural evolution. Confirming the precocious separation of reductive and expansive populations are significant psycholinguistic and ecological differences. One may therefore hypothesize that ethnic and environmental factors determined the fissure between the interventions' two forms.

BIBLIOGRAPHY

Battaglia R. Africa, Genti e Culture, cap. 2, Istituto Italiano per l'Africa, Roma,1954.

Baumann H and Westermann D. Les peuples et les civilisations de l'Afrique, Payot, Paris,1948.

Bruce J. Voyage au source du Nil en Nubie et en Abyssinie,1768-1773. Pancoucke, Paris, 1791.

Bleek WHI and Lloyd, LC. Specimens of Bushmen Folklore. C. Struik (PTY), Cape Town, 1968.

Cavalli-Sdorza F. Selezione naturale e selezione culturale. Il popolamento dell'Europa e la ricerca di un'identità, European Union Cross Identità Network, Univ. di Bologna, Convegno "Natura, Cultura, Appartenenza: identità culturali europee", 2002.

Cuvier G. Extraits d'observations faites sur le cadavere d'une femme connue a Paris et a Londres sous le nom du Venus Hottentotte, *Mèmoires du Museum d'Histoire Naturelle,* 3:259-266, 1817.

Dart RA. The physical characters of the Auni-Khomani Bushmen. *Bantu Studies*,11:176-246, 1937.

De Rakewiltz B. Eros Nero, costumi sessuali in Africa: dalla preistoria ad oggi. Longanesi e C., Milano, 1963.

Erlich M. La femme blessée: essai sur les mutilations sexuelles feminines, L'Harmattan, Paris, 1986.

Fucaschi M. I segni sul corpo. Per un'antropologia delle modificazioni dei genitali femminili. Bollati Boringhieri, Torino, 2003.

Grassivaro Gallo P. La circoncisione femminile in Somalia. Una ricerca sul campo. F.Angeli, Milano, 1986.

Grassivaro Gallo P. Figlie d'Africa mutilate. Indagini epidemiologiche sull'escissione in Italia, L'Harmattan-Italia,Torino, 1998.

Grassivaro Gallo and Viviani F. The origin of infibulation in Somalia: an ethological hypothesis. *Ethology and Sociobiology*,13:253-265, 1992.

Grassivaro Gallo P. and Villa E. Longininfismo rituale tra le Baganda (Uganda). Rapporto Preliminare, *Rivista di Sessuologia*, 28/1:17-22, 2004.

Grottanelli VL. Gerarchie etniche e conflitto culturale, F. Angeli ed., 1976.

Gruenbaum E. The female circumcision controversy. An anthropological perspective. Penn,Univ. of Pennsylvania Press, Philadelphia, 2001.

Guth D and Von Samsonov E. Sex politik, Lust zwischen restriktion und subversion, Hrsa.vlg. TURIA-KANT,Wien, 2001.

Hicks EK. Infibulation status through mutilation. Tesi di specializzazione,

Faculteit der Sociale Wetenschappen, Erasmus Universiteit, Rotterdam,1986.
Hosken FP. The Hosken Report. Genital and Sexual Mutilations of Females. WIN News, 2nd & 3rd eds., Lexington (Ma),1982,1994.
Jeffcoate TNA. Principles of Gynecology. Butterworth and Co. Publishers, 3th ed., London, 1967.
Kasamura A. Famille, Sexualité et Culture. Essai sur les moeurs sexuelles et les cultures des peuples des Grandes Lacs Africains, Payot, Paris, 1973.
Le Vaillant F. Le voyage du Monsieur Le Vaillant dans l'interieur de l'Afrique par le Cap de Bonne Esperance, dans les années 1780,1,2,3,4, et 5, Paris, Leroy,1790.
Livibacci M. Introduzione alla demografia. Loescher, Torino,1990.
Makie G. Female Genital Cutting: the beginning of the end, pp. 252-281. In: Shell-Duncan B. and Hernlund Y. (Eds.) Female "Circumcision" in Africa. Culture, Controversy and Change. Lynne Rienner Publishers. Boulder (Co), 2000.
Peron P and Lesueur CA. Observations sur le tablier des femmes Hottentottes. *Bull. de la Soc. Zoologique de France*, 8:15-83,1883.
Puccioni N. Delle deformazioni e mutilazioni artificiali etniche più in uso, Archivio per l'Antropologia e l'Etnologia, 34:391-401,1904.
Rugiu G. I Boscimani. *Rivista di Antropologia*, 29:425-504, 1931.
Schapera I. The Khoisan people of South Africa. Routledge et Keagan. London, 1930.
Schultze-Jena L. Zur kentniss des Korpers der Hottentotten und Buschmanner, Jenaische Denkschriften, 17:147-227, 1928.
Seligman CG. Races of Africa. Thornton Butterworth. London, 1930.
Shell-Duncan B and Hernlund Y. Female Circumcision in Africa: dimensions of the practice and debates, pp. 1-39. In: Female "Circumcision" in Africa. Culture, Controversy and Change. L. Rienner Publ., Boulder, Co, 2000.
Sonnini de Mannoncour CS. Voyage dans la Haute et Basse Egypte fait par ordre del'ancien Gouvernement, F. Buisson, Paris, 1798.
Stow GW. The native races of South Africa. Swan Sonnenschein & Co, London, 1905.
Taha Baasher 1982 Psychosocial aspects of female circumcision, pp. 162-180. In: Taha Baasher, Bannermann R.H.O., Rushwan H., Sharaf I. (Eds.) Traditional practices affecting the health of women and children. WHO/EMRO Technical Publications, no. 2, vol. 2, Alexandria.
WHO Female Genital Mutilation. Report of a WHO Technical Working Group, 17-19 July, 1995, Geneva, 1996.

Chapter 5

PSYCHOLINGUISTIC APPROACHES TO RITUAL LABIA MINORA ELONGATION AMONG THE BAGANDA WOMEN OF UGANDA

Elisabetta Villa and Pia Grassivaro Gallo
Padua Working Group on Female Genital Mutilation, Department of General Psychology. University of Padua (Italy)

Abstract: Ritual elongation of the labia minora is a unique enlarging modification of the external genitalia exercised for cultural reasons (FGM type 4 - WHO 1996). The practice is common among the Baganda women of Uganda, where a variety of terms are used to describe the rite. Psycholinguistic analysis was conducted both in present day Africa, where elongation of the labia minora results from ritual manipulation, and through the bibliographical accounts of western authors (anthropologists and doctors) from the 1950s and 1960s. A semantic polarization results in the linguistic expressions. In Africa, the positive connotation of terms used to describe the rite indicates its substantial valorization. The vocabulary used by Western authors, however, includes reference to aspects of rural Europe suggestive of poverty and ignorance ("apron"), or symbolic ridicule of the manipulated feature, equating it to the ear of a Cocker Spaniel.

1. PREFACE

As language is an instrument that is fundamental to the life of a community, it is an element impossible to ignore when considering daily exchanges. Especially for adults, all features of social life are expressed through language: ceremonies, prayer, oaths; and, through oral tradition, retellings of past events that allow the memory of a community to be

G.C. Denniston et al. (eds.), Bodily Integrity and the Politics of Circumcision, 57—64.

preserved (Cardona, 1976). In recent years, anthropologists fascinated by this condition have shown increasing interest in developing a deeper understanding of non-European languages that, for the enrichment that they provide, are considered carriers of implications and values connected to a conceptualization different from our own; thus creating the field of ethnolinguistics.

In traditional African society, the elderly were the keepers of knowledge given the task of transmitting essential morality through oral tradition. Among the populations of Uganda, this was conveyed by considerable use of proverbs, sayings, maxims, riddles, epic poems, and fables. The expertise of an elder was not limited to telling, but rather shown in knowledge about how to apply such proverbs to different situations, offering suggestions rather than solutions, and leaving it up to the listener to find an answer.

Education of children in village life also took place through words and colloquial expression rich with allusions, plays on words, and double meanings. The content results are complex and comprehensive, and the language is well described as poetic (Cisternino, 1987).

As for ritual elongation of the labia minora (WHO female genital mutilation type 4, 1996), a rite by which the body of an adolescent girl is prepared for marriage, the *ssenga* was the traditional person trusted with the upbringing and sexual education of her niece until marriage. This role was carried out, most especially, through words and practical demonstration (Grassivaro Gallo & Villa, 2004).

2. PSYCHOLINGUISTICS OF ETHNIC FEMALE GENITAL MODIFICATION (EFGM)

According to Cardona (1976), the topography of the human body serves as a model on which a variety of linguistic realities are projected. In several known examples (Friedrich, 1969), every word denoting a body part, in addition to its function in mapping the body, has a secondary meaning regarding the genital area. In contraposition to these two possibilities, indicating the strictly physical body, stands reference to the external world (the house, trees, etc.), distinguished from the former, and expansive into the psychosocial sphere. It is not to be said that such an expansion is universal, however, where present, this proves useful in revealing important features of human behavior that transpire from terms used and their attributed significance (Taha Baasher, 1982). Considering this, the psycholinguistics of ethnic female genital modification is of great interest.

Among the African diaspora, for example, female immigrants from the zones of Africa where excision is practiced are deeply offended by Western use of the term "mutilation" to indicate interventions on the external genitals. Therefore, we suggest the adoption of the neutral term "ethnic female genital modification" (Puccioni, 1909; Fusaschi, 2003), which is more acceptable to all.

Another relevant psycholinguistic consideration noted in the past (Taha Baasher, 1892; De Rachelwitz, 1963) is reconfirmed through an analysis of the elongation of the labia minora in Uganda. The vocabulary used to indicate all expressions inherent to modifications of the genitals, anatomical terms, and the associated customs and behaviors, when not completely absent, is extremely limited in populations that excise the genitals ("reductors") (Fusaschi, 2003). Contrarily, populations practicing modifications that increase the size of the genitals ("expansors") have a considerable vocabulary at their disposition that is rich with differentiated linguistic structures for the discussion of sexual topics (Kashamura, 1973).

For instance, in the Sudan, the terms used to indicate infibulation are limited to three nouns: *khitàn*, meaning "circumcision;" *khifàd* (from the Arabic word *khafad*), "to reduce;" and *taòur* (from the Arabic *tahaar*), "to purify." These are always followed by the adjective *fàraòni* (pharonic) (Toubia, 1999; Rizzo, 2001-2002).

In Somalia, the most general term used to indicate female circumcision is *gùdnìn*. The two possibilities are *sunna gùdnìn* (attenuated circumcision), and *cadi gùdnìn* "the normal," to designate" infibulation."

Among "reductor" populations, there is an apparent absence of a term for masturbation. This is the case in Somali (Omar Abdulkadir, personal communication, 2003), in Eritrea (M. Cosentino, personal communication, 2003), and in Egypt, where the idea is conveyed through a euphemistic expression constructed from the Arabic words for "habit" and "secret;" *el ada el serreya* (Romagnoli, 2003).

In Somalia, neither the concept nor the word "virginity," meaning "integrity of the hymen," exist. Infibulation instead serves to create the desired physical condition (Mohamed Somalia, circumcision operator in Mogadishu, personal communication, 2004). In the Sudan, the expression "breaking virginity" refers to the first marital intercourse with infibulated women (Hamid Al-Bashir Ibrahim, 1997).

The situation is reversed in "expansor" populations. According to Kashamura (1973), "... in all of the Great Lakes region (of Africa), where the vulva is artificially elongated ... a significant vocabulary, rich with metaphors used in discussion of sexual issues is available. For example, there are about ten different ways to indicate the female sexual organs."

3. PSYCHOLINGUISTICS OF LABIA MINORA ELONGATION AMONG THE BAGANDA OF UGANDA

It is important to note that all interviews about genital elongation were conducted in the local language, *Luganda*, even though our subjects (school girls, teachers, and wise women who execute the genital manipulations) generally spoke English.

References to the practice are common in women's conversation, as it is held to be fundamental. The subject, however, is not discussed between mother and daughter, as it is considered taboo (Wele, 1993; Francheschini, unpublished data).

Genital manipulation is indicated by a variety of terms. The most common, *okukyàlira ensìko* (visiting the bush), refers to the clearing where, safe from view, the first intervention will take place at the hands of the *ssenga* (paternal aunt), or occasionally the *jaja* (a grandmother figure, who in the extended African family assumes a more important role than her Western counterpart).

A more common expression is quite graphic: *okusìka enfùli* literally means "pull the vulva." The elongated labia minora are called *mfuli* (the same term in Swahili is used by Luba and Nkundo women of the Congo, by the Hottentot's of South Africa, and by the Swahili of Tanzania and Kenya (De Rachewiltz, 1963). Gisu women of Uganda also use the term *mfuli* to indicate the manipulated labia minora, which assume "the morphology of a turkey's crest" (De Rachewiltz, 1963).

The elongate labia are also called *abalòngo* (twins) because the two must assume equal length for symmetry, and because their creation is believed to favor the birth of twins. Such an event, always of great auspice in a multiparous culture, is especially esteemed among the Baganda, where it is retained that a couple having produced twins is blessed with particular capacity to influence the growth of bananas, a local food staple. During the initiation of twins, the parents are encouraged to engage in sexual intercourse (Kisekka, 1973), which is enhanced by the practice of labial elongation (*okusikina*), as it increases the reciprocal pleasure of both partners (Sengendo and Sekatawa, 1999).

A celebration that is both sacred and profane follows the birth of twins. The father, using his penis, knocks a banana flower out of the hands (or the vagina) of the mother, who lies on her back in the grass. The couple then dance in their friends' orchards to insure an abundant harvest (Fraser, 1922; De Rachelwiltz, 1963).

The enlarged labia are also called "doors" (a recent Anglo-Saxon expression); the significance of which was described by a Baganda wise woman, "building a house would not be possible without doors to let you in," and by another, "...passing through the door, you are obliged to respect the house in which you enter." A Ugandan proverb says "*bulugadde batya* (*oluggi*)" (how could they make such a narrow door? We can't get anything through (Walsen, 1982). A girl who undergoes the rite will no longer be an *omugaso* or *kyangaala* "something completely open or unadorned" as though she had a hole; she will also be spared from disparaging expressions, such as *kiwompogoma, kiwowongole, kifufunkuli, funkuli muwompogoma.* If she does not undergo the rite, she will remain forever *odiba* (literally, "not ready"), being a girl who must remain unmarried. On the other hand, "after having pulled, one grows rapidly, and the boys begin to arrive." "During intercourse, a man will ask the woman if he can touch her labia minora. After having received permission, he will softly pinch, assuring that his partner feels no pain during these acts" (Sengendo and Sekatawa, 1999). "This pull is not bad because it helps a mother when she must give birth, in fact, it enlarges the path for the baby."

A true Baganda woman with enlarged labia will become "more receptive and hot." This is a sign of respect and welcoming for her future husband, as it "helps to pleasure while one plays at sex." "Sex, in fact, is considered a game, and "playing sex" is the expression most used to indicate coitus (Forster and Furley, 1989).

The plants used by the *ssenga* in the first intervention are also denoted through highly symbolic names: *namirembe* (literally, "bringer of peace"), being auspicious for conjugal peace and stability; *kabbo ka bakyala* (literally, "women's cup or basket"); *mukasa*, name of the divinity of Lake Victoria. At the birth of a son or of twins, she is always thanked for bounty and abundance. In the past, a girl who did not want to undergo the rite would be thrown into the same lake.

Finally, it is interesting to examine the word used to denote the handkerchief that the *ssenga* uses to dry her hands after the first manipulation of her niece. In Lugandan, it is called *nkumbi* (garden "hoe"). The symbolic significance is clear, as the hoe serves to till the earth in preparation for an eventual harvest, the rite of labia minora elongation prepares the girl's body for matrimonial fertility.

In Western literature, however, we find that English anthropologists and geographers have used the term "apron" to describe the elongated labia minora of Hottentot women (Stow, 1905; Scapera, 1939); French anthropologists have called it a *tablier* (Le Vaillant, 1790), and Italian academics have called it a *grembiule* (Rugiu, 1931). These related terms refer to the morphology of the altered parts, which reach as far down as the middle of the thigh, but these terms also refer to an article of clothing characteristic of rural Europe.

The observed morphology assumed either the appearance of a "folded butterfly" wing in the tribes of South-West Africa: in the Naron and Auen of Kalahari, or, less expansive at the attachment and more so in the exterior margin, of a "turkey wattle" in the Bushmen of the Province of the Cape and of Botswana (Drury and Drennman, 1926; Dart, 1937; Baker, 1974).

In a book on gynecology, we have found a unique expression for the elongated labia minora that describes them as "spaniel ear *nymphae*" (Jeffcoate, 1967). We have also found terms derived from obsolete anatomical vocabulary: *longynymph* (Drury and Drennam, 1926), translated in Italian as *longininfismo* (Battaglia, 1954), and in French as *macronymphie* (Erlich, 1896; Fauvelle-Aymar, 2002); deriving from the term *nymphae*, which designates both the labia minora and labia majora.

4. COMMENTS

From the psycholinguistic analysis of the various expressions used in Africa and used by western authors, one can make several observations. First, in both the African and western vocabularies, some terms refer to anatomy; *longynymphism, mfuli*, etc. These are of less interest from a psycholinguistic perspective. Among the remaining expressions, a semantic polarization is evident.

In Baganda vocabulary, a substantial valorization of the local tradition emerges from repeated reference to tools and important phases of agricultural life: twins, with ceremonies for the fertility of fields; the door, an indispensable element of the house to protect the family; the forest, a sacred place where the first intervention by the *ssenga* occurs (similar to the *bois sacré*- Kashamura, 1973); and the hoe, which makes the earth fertile. The names of plants used in genital manipulation are also given positive connotations, referring consistently to fertility gods. Completing these observations is the respective linguistic valorization or degradation of women who have or have not accepted the practice.

Expressions used by Western authors refer to a daily aspect of the rural European environment; the apron. This was probably connected to the poverty and ignorance that often coexisted in lower class populations. Other terminology ridiculed the genital morphology, comparing it to the ear of a cocker spaniel. Such references, used primarily during the middle of last century, indicate a disparaging view of the African, condemning his traditions because they are different from our own.

5. CONCLUSION

The psycholinguistics of female genital modification reveal a reductive/ expansive dichotomy that is reflected in the sexual customs of the populations examined. A vast gulf divides the "expansors," who are generally more comfortable with sexuality and who possess a rich and specific vocabulary on the subject, from the "reductors" (excisors and infibulators), who have a negative view of sex and whose vocabulary on the subject is relatively poor.

BIBLIOGRAPHY

Baker JR. Race. Oxford Univ. Press., London, 1974.

Battaglia R. Africa, Genti e Culture. Istit.Ital. per l'Africa. Roma, 1954.

Cardona GR. Introduzione all'Etnolinguistica. Il Mulino, Bologna, 1976.

Cisternino M. Ape segreta fa miele profumato. Nigrizia, 105, giugno:51-53, 1987. (ext. ed.): The Proverbs of Kigezi and Ankole (Uganda), Museum Combonianum, 4, Kampala/ Roma, 1987.

Dart RA. The Phisical Characters of the Auni/Khomani Bushmen, Bantu studies, 11:176-246, 1937.

De Rachewiltz B. Eros Nero. Longanesi e G., Milano, 1963.

Drury J and Drennman MR. The pudendal parts of the South-African Bush Race, Medical J. South Africa, 22:113-117, 1926.

Erlich M. La femme blessée: essai sur les mutilations sexuelles femminines. L'Harmattan Ed., Paris, 1986.

Fauvelle-Aymar FX. L'intion du Hottentot. Histoire du regard occidental sul les Khoisan (XV-XIXe siècle), publ. De la Sorbonne, Paris 2002.

Forster SJ and Furley KE. Public Awareness Survey on AIDS, 3(3):153-156, 1988.

Franceschini PL. (a cura di) Appunti sul Malawi, pp. 75-76, unpublished data.

Fraser JG. Il ramo d'oro. Newton Compton Ed., Roma, 1922.

Friedrich P. On the Meaning of the Tarascan Suffixes of Space, Internat. J. American Linguistics, memoir 23, Bloomington, IN, 1969.

Fusaschi M. I segni sul corpo. Per un'antropologia delle modificazioni dei genitali femminili. Bollati Boringhieri, 2003.

Grasivaro Gallo P and Villa E. Longininfismo rituale tra le Baganda. (Uganda) (RapportoPreliminare). Rivista di Sessuologia, 28:17-22, 2004.

Hamid Al-Bashir Ibrahim, (original edition, arabic language) 1997. La donna e la triste realtà culturale sudanese. Ricerca per capire l'ideologia dell'infibulazione delle ragazze in Sudan.; www.islamitalia.it, 2002.

Kashamura A. Famille, Sexualitè et Culture. chapt. 3°, La Vie Sexuelle, pp. 81-123, Payot, Paris, 1973.

Kisekka MN. The Baganda in Central Uganda, pp. 148-162. In: Molnos A. (Ed.)

Central Source Materials for Population Planning in East Africa, vol. 3, Beliefs and Practices, East Africa Publishing House, Nairobi, 1973.

JeffocateTNA. Principles of Gynaecology, 3th. ed., Butterworths and Co. Publishers. University Press, Aberdeen (UK), 1967.

Le Vaillant F. Voiage de Monsieur Le Vaillant dans l'interieur de l'Afrique par le Cap de Bonne-Esperance, dans les années 1780, 81, 82, 83, 84, et 85, Paris, Le Roi, 1790.
Puccioni N. Delle mutilazioni e deformazioni artificiali etniche più in uso. Arch. per l'Antrop. e la Etnol., 34:391-401, 1904.
Rizzo T. La circoncisione femminile in Sudan. (Un'esperienza con studentesse sudanesi),Tesi di Laurea in Psicologia, University di Padova, AA.2001/02.
Romagnoli G. Se il piacere è muto. La Repubblica delle Donne, p. 44, 1/3/2003.
Rugiu G. Rivista di Antropologia, 29:425-504,1931.
Schapera I. A Survey of the Bushman Question, Race relations, 6:68-83, 1939.
Sengendo J and Sekatawa EK. A Cultural Approach to HIV/AIDS Prevention and Care. UNESCO/UNAIDS Res. Proj. Uganda's Experience Country Report.
Studies and Reports, Special Series, no. 1, UNESCO, 1999.
Stow GW. The native races of South-Africa. Swan Sonnenschein & Co.; London, 1905.
Taha Baasher Psychosocial Aspects of Female Circumcision, pp. 162-180.
In: Taha Baasher, Bannermann RHO, Rushwan H, Sharaf I. (Eds.). Traditional Practices Affecting the Health of Women and Children. WHO/EMRO Technical Publications, no. 2, vol. 2, Alexandria, 1982.
Toubia N. Caring for Women with Circumcision. Rainbo.
Venus, circle over a cross), New York, 1999.
Walsen F. Luganda Proverbes. Mill/Hills Missionaries, London, 1982.
Wele P. "Likumbi Lya Mize" and Other Luvale Ceremonies. Zambia Educational Publishing House, Lusaka, Zambia, 1993.

Chapter 6

GRAPHIC REPRODUCTION OF GENITAL STRETCHING IN A GROUP OF BAGANDA GIRLS

Their Psychological Experiences

Pia Grassivaro Gallo, Elisabetta Villa, Fabiola Pagani
Padua Working Group on Female Genital Mutilation, Department of General Psychology, University of Padua (Italy)

Abstract: In 2002, Padua's Working Group on FGM organized a mission to southern Uganda to analyze the rite of labia minora elongation among the Baganda. This manipulation has been classified among the less common forms of FGM (type 4) by the WHO (1996). The practice occasionally occurs in West African populations, where it is referred to as hypertrophy of the labia minora. The psychological experiences of the rite were emphasized by the analysis of drawings and the comments on the rite, made by 111 schoolgirls, from 12 to 16 years old; all of the girls examined underwent genital stretching. The results highlighted two different groups of subjects: those who adhere to and those who criticize and are somewhat opposed to traditional stretching. These results are confirmed by comparison of the drawings of female circumcision made by Somali schoolgirls of the same ages, studied with the same methodology.

1. INTRODUCTION

In 1996, the World Health Organization (WHO) proposed a classification of Type 4 Female Genital Mutilation (FGM) dedicated to the less common manipulations, including the elongation of the clitoris and the labia.

In 1998, the same organization listed Uganda among the nations in which FGM is endemic, citing a prevalence of 5% (513,050 mutilated women belonging to two ethnic groups out of a population of ten million female

G.C. Denniston et al. (eds.), Bodily Integrity and the Politics of Circumcision, 65—84.

inhabitants). A footnote addressed the anecdotal nature of the data, as no field study was available at the time.

In 1999, the WHO described Project REACH (Reproductive Education and Community Health Organization) in great detail, as aimed at the eradication of excision among the Sebei people of the Kapchorwa District in southeast Uganda. This appears to be the only form of FGM observed in the country.

In the summer of 2002, the University of Padua Working Group on FGM organized a mission among the Baganda of southern Uganda to carry out an analysis of the ritual by which the labia minora are elongated. The undertaking was developed on the recommendation of two ethnically Baganda missionary Verona Fathers who lived in Italy temporarily. A third Verona Father, of Italian descent, A. Dalfovo, an anthropologist at Makerere University of Kampala, gave directives for the selection of subjects, to whom the graphic tests were administered.

During the mission, data of an anthropological, as well as psychological nature, were collected. Only the latter are presented in this paper.

2. GENITAL STRETCHING AND OKUKYALIRA ENSYKO IN UGANDA

The labia minora, two cutaneous folds interior to the labia majora, are part of the female external genitalia. Histologically, they present ample vascularization in the connective tissue below the mucosa because they behave like erectile cavernous tissue (Balboni *et al.*, 2000). Among their characteristics, there is a high concentration of sensory nerves (WHO, 1998), so that, according to Lowry (1976), "the labia minora and clitoris are primary sensory organs in female sexuality, therefore, mutilation of even one of these parts interferes in the receptivity of female sexual stimulation."

On the other hand, manipulation can increase neurosensitivity in the area, resulting in increased sensibility of women with hypertrophic labia who react more intensely to genital stimulation (Dikinson). This ritual stretching is performed by many populations in the eastern regions of central southern Africa (Grassivaro Gallo and Villa, 2004; in press).

Labia minora elongation has been identified as an exotic, racial characteristic by European anthropologists and researchers in systematic treatments or descriptions of the Bushmen/Hottentot's (Baumann and Westermann, 1948; Battaglia, 1954; Biasutti, 1959), rather than being recorded among the biological aspects of female sexuality by psychologists/sexologists or cultural anthropologists (Bettlelheim, 1971; Van Gennep, 1909; De Rakewiltz, 1963).

By the 1960s, the hypertrophy of the labia minora was recognized among European populations in obstetric/gynecological conditions, listed among the pathologies of the labia minora (Jeffcoate, 1967), and described using the unusual phrase “spaniel ear nymphae.” Either mono or bilateral, the condition was said to emerge during puberty, resulting from poor hygiene, chronic irritation, or excessive masturbation; sporadically they appear as a spontaneous case, due to the ample variability in the dimensions of the labia minora between individual subjects. The accentuated sensitivity connected to hypertrophy can sometimes result in discomfort. Such a disturbance can be eliminated with reduction by labiaplasty, considered among the aesthetic surgical operations that render the body more acceptable by attempting to counteract the natural evolution of age and/or manipulations (masturbation and/or sexual intercourse, in this case). With such an operation, the superfluous labia are drastically reduced in order to permit direct vaginal penetration (Guth and Von Samsonov, 2001).

The ritual stretching among the Baganda women (*okulyalyra ensyko* in luganda and “visiting the bush”) from the anthropological data was described for the first time by Grassivaro Gallo and Villa (2004; in press).

The stretching takes place in a forest clearing, when the operator (called the *ssenga*, usually a paternal aunt) instructs her niece to manipulate her genitals and to gather specific herbs; the first operation (being the most painful) occurs when the aunt proceeds with the first manipulation, aided by the juices of selected fruits and herbs. The teaching phase lasts several years, during which the girl will practice auto-manipulation in more private settings until the labia minora have achieved the proper length of ten or more centimeters.

Changes to the rite in Uganda have followed urbanization, disintegration of the extended family, and school education of adolescents. The practice still exists among diaspora populations as a sign of respect for tradition.

The stretching of the labia minora aims to increase pleasure for both partners during sexual intercourse and thereby create a large family. The husband’s satisfaction reflexively gratifies the woman, who enjoys an enhanced social status.

3. SAMPLE AND METHODS

In Uganda, 111 Baganda girls, twelve to sixteen years (mean age, 13.9, SD = 1.1), participated in this study. All the girls underwent a similar experience of labia minora stretching; the mean age at the beginning of the rite was 10.8.± 1.5; the mean difference between the above quoted ages was 3.0. ± 1.8 years. No correlation emerged between these ages ($r = -.07$; $p < .42$). Data were collected in the schools of Nkokonjeru, a village near Kampala.

With the help of two interpreters, a set of graphic tests was handed out to each class. The tests used were the Tree Test (Koch, 1957); the Human Figure Test, in two versions (Machover, 1949); and the Family Test (Corman, 1967). In addition, the subjects were invited to draw the rite and then write explanatory comments.

In this study, we shall examine only the last of the five tests. The rest of the drawings were used as a reference to determine the graphic skills of the subjects. Both the drawings and comments on the stretching rite were analyzed on the psychological level, that is, what the experience meant for the subjects. For the data coding, each drawing was considered a case in itself. A number of qualitative characteristics related to form and content, aspects of each drawing were classified; then other quantitative variables were added (such as actual age, ritual age, etc.). The analysis of the drawing focused on elements involving environment, subject, and traditional operator (*ssenga*). A first analysis was conducted through the raw percent of each characteristic.

The characteristics, recovered from both the qualitative and quantitative examinations of the drawings (such as environment, setting, operator, and subject representation), were classified on the basis of either presence or non-presence, for each subject. We considered only the ones that were present on drawings with a frequency not less than ten percent, in order to prevent rare features from attaining prominence.

The statistically significant characteristics of the drawings were submitted to an agglomerative hierarchical cluster analysis using Ward's method (Aldenderfer and Blashfield, 1984; Fabris, 1997), which is based on squared Euclidian distances, to assess the distance among the clusters.

To determine the optimal number of the cluster, we examined the dendrogram in order to locate large increases in rescaled distances that would indicate good clustering solutions for the number of clusters immediately preceding the increase. The outcome showed that a two-cluster solution seemed to be the fittest to represent the data.

To validate a clustering solution, we performed significance tests (X2) on qualitative variables used to create clusters and significance tests (Student t), which compare the clusters on the quantitative variables not used to generate the cluster solution.

Finally, using some indexes (as recollection of pain, happiness, social and/or affective presence), we also were able to analyze the Baganda girls' written comments; they are divided in two groups: depicted spontaneously by 20.7% of the subjects; not spontaneously for the others because two interpreters specifically requested the answers during the administration of the graphic tests.

The graphic productions of the Baganda girls were compared with the drawings and comments on female circumcision made by 93 Somali girls of Mogadishu (Grassivaro Gallo and Moro Boscolo, 1984-1985).

Their mean age was 13.2, SD = 1.1; the mean age at circumcision was 6.1, SD = 1.9; the mean difference between the above quoted ages was 7.1, SD = 2.0.

The same methods were used also for Somali graphic drawings and comments.

4. RESULTS

4.1 Uganda

Above all, the drawings depicting the rite represent a graphic involution compared to the other test drawings. This characteristic concerns the outline of the human body and the bush environment (respectively, 43.2 and 20.7%). The figures are frequently crippled (43.2%).

The moment of the rite, chosen for representation, is an auto-manipulation by the subject or the *ssenga*'s first intervention, otherwise the rite is censured and the subjects produce only minimalist, schematic, and static figures (respectively, 53.1; 12.6; 27.0%).

Usually the drawing focuses on the subject's figure (65.8%). The forest environment is always present (Figure 1); her home only rarely (Figure 2) (respectively, 91.9; 6.3%).

Figure 6-1. The Stretching Scene. U-13 years old.

It is interesting to analyze the typology of the single figures that appear most frequently: the subject is always present, drawn alone or together with others figures (respectively, 65.8; 34.2%); frequently the body is depicted as crippled; she appears bare with highlighted genitalia, and her face expression is indistinct (respectively, 27.9; 52.2; 81.1%) (Figures 2, 3, 4).

Figure 6-2. The stretching scene. U-14 years old.

Figure 6-3. The subject in auto-manipulation.

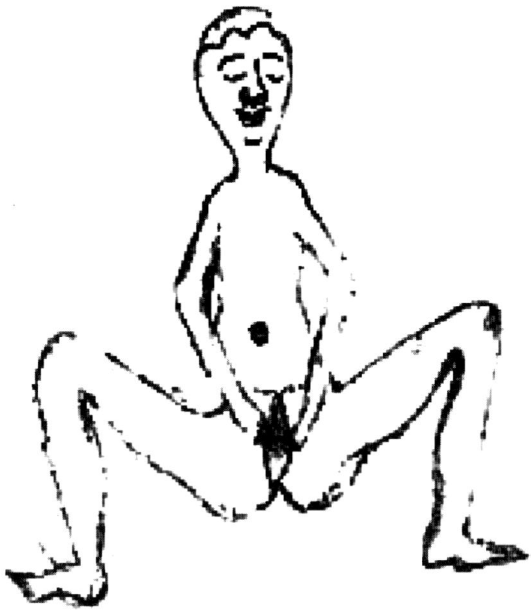

Figure 6-4. The subject in auto-manipulation.

The operator (*ssenga*) is rarely depicted (21.6%); her face appears inexpressive, often with a crippled body and clad in the traditional style (respectively, 87.5; 45.6; 37.5 %) (Figures 1, 5, 10, 11).

Figure 6-7. The operator manipulating vulva of the girl. U-16 years old.

Cluster analysis yielded two very distinct groups, each of them containing drawings characterized by a different meaning given to genital stretching by the Uganda girls (Dendrogram U).

- Cluster 1 groups 59 girls (from number 69 to 27); the mean age was 14.1, SD = 1.1; the mean age at the beginning of the rite is 10.6, SD = 1.6; the mean difference between the above quoted ages was 3.5, SD = 2.0.

- Cluster 2 classifies 52 girls (from numbers 64 to 30); the mean age was 13.5, SD = 1.1; the mean age at the beginning of the rite was 11.1, SD = 1.3; the mean difference between the above quoted ages was 2.5, SD = 1.2.

Validation techniques of the clusters:

- The significance of the (X2) test on variables used to create clusters results in the following as discriminates: amputation ($<.01$), transparencies ($p < .05$), hair ($p < .01$), arms ($p < .01$), hands ($p < .01$), feet ($p < .01$), and genitalia ($p < .01$). The first variable (amputations) characterizes Group 1 (Graphic U) as the only discriminant.

- The significance of the test on the external variables to create clusters results in the following: the represented ritual moment is clearly depicted ($p < .01$); the subjects are represented auto-manipulating themselves ($p < .01$), either alone or together. Among the particulars, legs are often spread and well depicted. In cluster 2, the moment when the rite began is indistinct and, therefore, the figure of the subject is depicted with amputated legs, and in no instance are the legs wide apart.

Two clusters in comparison:

Cluster 1: the older subjects ($p < .01$) start the rite more precociously; they are the nearest to the auto-manipulation phase and thus to the end of the rite ($p < .01$). They depicted themselves in that phase with a figure complete with many body details. (Figures: 1, 3, 4, 5, 6, 13, 14).

Cluster 2: the subjects are younger, the beginning of the rite had been delayed, and they were probably still incomplete at the time of our interview. They often depicted themselves in a minimalist way and with crippled bodies. (Figures: 2, 7, 8, 9, 11, 12).

In the spontaneous comments, mostly written by Cluster 1 girls, they always refer to physical pain, especially that of the first operation:

> It's a painful rite, especially when it's the first time you do it. You feel pain at the beginning but at the end it's painless.

Completing the social meaning of the rite, the connotations of "good" or "important" are associated to the adjective "painful":

> It's good, although it's very painful.

The importance is always focused on its reproductive meaning:

> They say that if you don't do it, you won't marry. It's done because it makes you "hot," and they say that men prefer it. Our aunts and mothers told us that, if you refuse it, you won't it able to give birth. People say that, if a woman doesn't do it, she isn't accepted as a woman.

It can be inferred that the opinions expressed are less those of the subjects than of others.

> Old people say that it's good.

Furthermore, the way that the girls graphically depict the progress of the auto-manipulation is worthwhile stressing.

> One can begin at any age and practice it regularly, every day twice or thrice; you don't waste time because you do it in your free time; you must repeat the action of stretching every time you finish your menstruation.

4.2 Somalia

In the drawings, the circumcision scene takes place indoors (93.6%), at home/hospital (Figures 15, 16, 17). A specific phase of operation is chosen only by 19.4% (Figure 20) of the subjects; they usually represent the period following the intervention (64.5%), with the party and the convalescence (Figures 18, 19).

The subject normally is drawn lying on a bed or a table for the operation (90.3%); often appearing with inexpressive face, but frequently showing pain/fear (59.1; 32.3%), bare or clad, with either indistinct or western clothes (30.1; 48.4; 20.4%).

The girls remembered, in 24.7% of the cases, that the circumcision was performed by a man/doctor, more rarely, a traditional woman/*midgàn* (Figure 21) (87.0;13.0%); the facial expression was indistinct, but often with an aggressive appearance (Figures 17, 22) (65.2;30.4%). In a few of the cases, he/she was depicted clad with either western or medical clothes (Figure 15) (21.7; 8.7%); otherwise, he/she was indistinct (69.6%). He/she brandishes surgical instruments (scissors, little bags, and drugs) (60.9%) (Figures 15, 16, 20, 21, 22).

From the cluster analysis (Dendrogram S), the subjects are divided into two groups:

Major Cluster 1 groups 63 girls (from numbers 17 to 58), the mean age was 13.2, S.D. = 1.2; the mean circumcision age was 6.0, S.D. = 1.7; the mean difference between the above quoted ages was 7.2, S.D. = 1.8.

Cluster 2 classifies 30 subjects (from numbers 31 to 60); the mean age was 13.2, S.D. = 1.1; the mean circumcision age was 6.3, S.D. = 2.3; the mean difference between the above quoted ages was 7.9, S.D. = 2.4.

4.3 Validation Techniques of the Clusters

The significance of the X2 test on variables used to create clusters results in the following as discriminants, tools/drugs ($p < .01$); hair ($p < .01$); arms ($p <. 01$); hands ($p < .01$); feet ($p <. 01$); body/dress details ($p <. 01$); amputations ($p < .01$). In particular, the last variable (amputations) was present (90%) in Cluster 1 only (Graphic S).

The significance of tests on the external variables used to create clusters, results in the following as discriminants: legs wide apart or not and ostentation of the genitalia.

The circumciser figure appears in both groups, but it is not a discriminant. Most subjects in Cluster 1 remember the tools and the drugs used to perform the operation: many have amputated limbs (hands and feet); few subjects depicted their genitalia and they are the only ones in this group. (Figures 15, 16, 17, 22).

In Cluster 2, the subjects' representations are more complete, with hair and clothes details (buttons and belts) (Figures 23, 25, 26).

4.4 Two Clusters in Comparison

The two groups have the same personal characteristics, such as age and the age of circumcision; therefore, we suppose that the division among the subjects was based on differing unknown variables (family origin, circumcision typology, and so on).

In Uganda, the discriminatory element is still a more relevant presence of amputations, which in Somalia are associated with a different representation of the genitalia.

In her comments, the girl remembers that she was not informed about what was going to happen to her:

> In 1976, they took me to the hospital without telling me anything about it; they put me on the table. The circumcision day I didn't know; while I wanted to go to the party, they took me to Dikfer Hospital.

On the other hand, one girl affirms:

> That day, it was Saturday. The one who circumcised me was named Ibado. Many people came to my house, friends; and I was happy because I was always asking my family to be circumcised. I was eight years old.

The circumstances of the operation are detailed meticulously: date, day of the week, hour of the day (45.2%), even the name of the circumciser is remembered, praising his/her skill:

> Dr. Sahal circumcised me at Dikfer Hospital. The doctor who circumcised me was named Charles, now he is dead.

> I was circumcised by Dr. Mahamed in my home. He was a clever doctor, who knew how to circumcise well.

From the details remembered, the day is "unforgettable." But what is mostly remembered is the party organized for the occasion: in particular, the presence of kinfolk and friends and the gifts received (68.8; 24.7%).

Sometimes the experience focuses dramatically on the pain suffered:

> That day I felt the worst pain in the world. I, Sucaad, felt a pain one can't believe. I felt a pain; I thought I could die.

The pain notwithstanding, the girl may have a good recollection of that day:

> When they seamed me, I felt pain, but I was happy.

That day is remembered negatively only in few instances:

> When they told me about it, I was afraid. That day was the worst in my life.

5. DISCUSSION

5.1 Uganda

On the basis of the data we analyzed, a profile emerges for the older Baganda girl (Cluster 1) who, having begun the rite at a young age, has been caught by our survey in the final phase of her auto-manipulation. The *ssenga*'s teachings, the repeated and periodical auto-manipulations, the social environment of general positive evaluation of the rite, the opinions expressed by the women elders on this issue, are all that apparently determined the girl's acceptance of tradition. The figure of herself in the graphic test, drawn as complete, rich with details of the body and the clothes, confirms this view. In the culturally rich, spontaneous comments written by these girls under their drawings, the recollection of the pain suffered, associated with its social importance, is a common feature. We did not find different words connoting this pain in their sentences but only the indication that connects it with the first ritual operation performed by the *ssenga*; the girls often point out that the pain fades away, it seems it does not have the dramatic consequences that are referred to by Africans in diaspora (Weinberg *et al.*, 2004).

It is interesting that the subject never expresses her own opinion, but that of somebody else. Two points are stressed: first, social identity still has an important value today, acquired through the rite; second, the reproductive meaning attached to tradition as an indispensable premise to marriage. Contrary opinions are few, and consensus seems to be unanimous, even if never uttered as the subject's opinion, but always preceded by the phrase "they say that ..." Clearly, the girl is influenced by other people's opinion.

Less is understood about the youngest girls, those who are closer to the first ritual operation. Due to their young age and school environment, their comments were decidedly poorer. Graphic representations, however, were very informative. These girls generally portray themselves as depreciated and simplified. Possibly, pain is responsible for these responses because it is still felt. The *ssenga* has not yet imparted her teachings to this age group.

5.2 Somalia

The results from the analysis of the clusters reveal that the Somali girls also can be divided into two groups. These groups cannot be distinguished on the basis of any personal characteristics, such as age, age of circumcision, and the distance between the two. If we use the same criterion used for

Uganda, we can argue that, although the rite is different, one third of the Somali girls examined (Cluster 2) accepted circumcision, since they drew themselves in the graphic test with a figure complete with body details and clothing accessories (buttons, belt, etc.). Most of them (Cluster 1), however, seem to be very "critical" of the operation because they graphically remember, even after many years, the cutting tools and medicines. They often drew themselves with limbs amputated (hands and feet), lying on their backs with spread legs, and clearly showing their genitalia. These details are rather disquieting. It appears that the variables that distinguish the two clusters refer to unknown cultural contexts, such as different types of circumcision (infibulation or attenuated form); medicalization of the operation, the use of a local or general anesthesia; the presence of the family women who care for the circumcised girl during the convalescent period.

We could not find elements confirming the subdivision into two groups in the comments. The content seems to be equally focused either on the fact that the girl felt offended at having suffered the operation without warning and was excluded from the party because her circumcision wounds incapacitated her or on the social evaluation of the convalescent period influenced by the party with her friends. The Somali group is particular and specific: the experience of the operation is dramatically influenced by the pain suffered, which indicates a more cruel operation (most often infibulation), performed in an unhealthy environment and characterized by a long, painful convalescence.

6. CONCLUSIONS

In this survey, we started from a tradition of female genital modification, which is customarily considered an essential prerequisite of the adolescent girl's body, in order to prepare her for matrimony and motherhood (Grassivaro Gallo, *et al.*, 2004). These rites are still performed in Africa with different methods and typologies.

We chose to analyze the graphic and verbal reports of two groups of Somali and Ugandan girls of the same age, studied with the same test method: the drawing of the genital modification, accompanied by a verbal comment. The results of these adolescents have been analyzed by the same method.

We must point out that the two analyzed populations are culturally different with regard to geographic area (Sub-Saharan and central eastern Africa, respectively), ecology (herders and agriculturalists), typology of genital modification (prevailing infibulation stretching), and psycholinguistics in the specific sector of genitalia (reducers and expanders), concerning the variety of the words used.

Despite these important differences, the graphic reports made by the adolescents we surveyed show a common fundamental dichotomy: both the Somali and the Ugandan girls are divided into two groups: one we may call "traditionalist," which adheres to the rite, and a "critical" group, which is somewhat opposed to it.

We also have identified among the Ugandan group the point when a girl, grown older, accepts the rite, probably because of the mediating role of the teachings and sexual education transmitted by the traditional operator (the *ssenga*).

If our findings are accurate and not flawed by an excessive degree of ethnocentrism, we can generalize and hypothesize that two views currently exist among the African populations that still perform these rites: The traditionalist view accepts the body modification of the adolescents for reproductive reasons; and the view of those who oppose it for health and general welfare reasons. Will the latter group prevail and convince the former? We hope this will be the outcome of an evolution of female genital modifications, which both keeps intact the connected cultural aspects and is content with autonomy, according to times and modes specific to every ethnic group.

BIBLIOGRAPHY

Aldenderfer MS, Blashfield RK. Cluster Analysis. Beverly Hills: Sage Publications, 1984.

Balboni GC. Anatomia Umana, 3rd Ed., vol. 2. Apparato Genitale Femminile, Milano: EDI Ermes s.r.l, 2000.

Battaglia R. Africa. Gentie Culture. Roma: Istituto Italiano per l'Africa, 1954.

Baumann H, Westermann D. Les Peuples et les Civilisations de l'Afrique. (Trad. francaise), Paris: Payot, 1948.

Bettelheim B. Les Blessures Symboliques (Trad. francaise). NRF Ed. Gallimard, 1971: Symbolic Wounds. New York: The Free Press, 1954.

Biasutti R. Razze e Popoli della Terra. vol. 1, Torino: UTET, 1959.

Corman L. Il disegno della famiglia: test per bambini. Torino: Boringhieri, 1967.

De Rakewiltz B. Eros Nero. Costumi Sessuali in Africa dalla Preistoria ad oggi. Milano: Longanesi & c., 1963.

Dikinson in: De Rakewiltz B. Eros Nero. Costumi Sessuali in Africa dalla Preistoria ad oggi. Milano: Longanesi & c., 1963. p. 120.

Fabbris L. Statistica multivariata. Analisi esplorativa dei dati. Milano: McGraw-Hill, 1997.

Grassivaro Gallo P and Moro Boscolo E. Female circumcision in the graphic reproduction of a group of Somali girls. Cultural aspects and their psychological experiences. Psychopathologie Africaine 1984/85:20;165/190.

Grassivaro Gallo P and Villa E. Longininfismo rituale tra le Baganda (Uganda). Rapporto Preliminare. Rivista di Sessuologia 2004:28;17/22.

Grassivaro Gallo P and Villa E. Ritual labia minora elongation among the Baganda women of Uganda (in press).

Grassivaro Gallo P, Tita E, Viviani F. At the roots of ethnic female genital modifications (Preliminary Report). The 8th International Symposium on Circumcision and Human Rights. Syllabus of Abstracts, p. 9. 2-4 September 2004. Padua (Italy).

Guth D and von Samsonov E. Sex politik, Lust Zwischen Restriction und Subversion. Wien: Turia & Kant, 2001.

Jeffcoate TNA. Principles of Gynaecology. 3rd ed., London: Butterworth, 1967.

Koch K. Il reattivo dell'albero. Il disegno dell'albero come ausilio in psicodiagnostica. Firenze: Giunti Ed.,OS, 1949.

Lowry TP. Some issues in the histology of the clitoris. In: Lowry T.P. *et al.* The Clitoris. St. Louis: Warren H. Green, 1976.

Machover K. Il disegno della figura umana. OS. Firenze, 1949.

Van Gennep A. Les rites de passage. Paris: E. Nourry, 1909. I riti di passaggio. Torino: Ed. Universale Scientifica Boringhieri. (ital. trad.), 1987.

Weimberg L, Mayer Fuchs C, Brooke-Condom J, Leigh Hobbs J, Ortiz A. Supplemental pretrial brief in support of respondents' application for asylum, withholding of removal, and withholding of removal under the convention against torture. Documents for the assistant District Counsel. Boston (MA): U.S. Department of Justice Executive Office of Immigration Review, Immigration Court. 26 October 2004.

WHO. FGM. Report of the WHO Technical Working Group. 17-19 July 1995. Geneva, 1996.

WHO. FGM: An Overview. Geneva, 1998.

WHO. FGM Programmes to Date: What Works and What Doesn't. A review. Geneva, 1999.

Chapter 7

SURVEY ON THE EAST-AFRICAN FEMALE STUDENTS AT THE UNIVERSITY OF PADUA

Stella Lineri, Chiara Rauhe, Pia Grassivaro Gallo
Padua Working Group on Female Genital Mutilation, Department of General Psychology, University of Padua (Italy)

Abstract: In 2003, the University of Padua organized the E. Piscopia Cornaro outpatient obstetric clinic to assist female personnel. The Psychology Department's Working Group on Female Genital Mutilation (FGM), in collaboration with the Committee of Equal Opportunities, has initiated a project to distinguish sexually mutilated East-African female students who may be potential patients of the clinic. A health education program has been created to better inform all female students and to introduce mutilated students to the possibility of deinfibulation. Among the nine East-African students approached, four with African origins have never thought of reversing the operation. Those of mixed heritage are not mutilated. All nine exclude the possibility of a future child undergoing excision. Introductory conferences have been organized to promote greater understanding of the phenomenon of FGM.

INTRODUCTION

In 2003, the University of Padua organized the E. Piscopia Cornaro outpatient obstetric clinic through the Department of Gynecology and Human Reproduction in an effort to offer assistance to female personnel.

This provided an opportunity for the Padua Working Group on Female Genital Mutilation (FGM) of the Department of Psychology, in collaboration with the Committee for Equal Opportunity, to launch a new project to identify sexually mutilated East-African female students as potential patients of the clinic, to organize a health education program to increase general awareness, and to familiarize mutilated students with the possibility of deinfibulation.

G.C. Denniston et al. (eds.), Bodily Integrity and the Politics of Circumcision, 85—91.

An introductory conference was held, in collaboration with one of the women's residence halls, in an attempt to promote understanding of the phenomenon of FGM.

1. THE RESEARCH PLAN

The proposed survey forms a pilot study promoted by the University of Padua, the first institution to organize an outpatient obstetric clinic. With appropriate publicity, it could spread to other universities.

The delicate nature of the subject of the project created a number of problems, and the collection of information from the East-African female students immediately proved quite arduous.

Obtaining names of potential subjects directly from the university or connected institutions was impossible due to privacy issues. Upon enrollment, students are guaranteed that their personal information will be used only to support their academic careers.

A request for the opinion of the Bioethical Committee about how to proceed received no response. Equally fruitless was an official list of foreign students at the University of Padua; available female names of East-African origin were scarce and, in some cases, due to privacy issues or missing addresses, it was impossible to approach the girls. In others, embarrassment over the subject proved an obstacle against direct contact with the students and their availability for interview.

The most useful method for contacting the previously identified subjects (East-African female students at the university), therefore, was "word of mouth," passed along through acquaintances, friends, and family.

We succeeded in approaching nine girls, all of whom have since participated with interest and dedication, protected by guaranteed anonymity. One in particular has been especially enthusiastic, bringing her personal testimony to several discussions organized on the topic of FGM.

Data was collected through interviews, structured "ad hoc" in order to expand understanding of pre-determined categories: the area of generalities, those relative to knowledge and experience regarding FGM, sexual experience and birth, deinfibulation, opinions on the media's treatment of the issue, and the utility of conferences open to all students on the subject.

Analysis of the information collected from the semi-structured interviews was completed using a psychosocial perspective and a qualitative cultural approach aiming to form a universal significance of the phenomenon and its most representative aspects.

2. THE SAMPLE

The results obtained, shown in the Table (Appendix B), summarize the demographic data, life events, and opinions of the nine East-African students of the University of Padua, the average age of which is 29.2. Two come from Chad, three from Sudan, and four from Somalia. They have been in Italy for an average of over five years, although the period of residence in

Table 7-1. General characteristics of East-African female students at the University of Padua
(AA 2003/2004)
SUBJECTS (N = 9).

Variable	Modality	Subjects (N.)
Age (in years)	Under 30	4
	31 e +	5
Origin of parents	Euro/African	4
	East-African	5
Country of origin	Chad	2
	Sudan	3
	Somalia	4
Nationality	East-African	5
	Euro/African	4
Permanence in Italy (in years)	Fino ad 1	3
	5 e +	5
	RND	1
Diploma	University	2
	Middle or high school	7
Civil status	Nubile	8
	Married with a child	1
Clothing	Traditional	3
	Western	6
Circumcision type	Infibulation	4
	Not circumcised	5

Average age (in years): 29.2 ±
Ds = 0.65
Average stay in Italy: 4.8 ±
Ds = 0.52
RND = No Response

three cases was only one year. Four have a double nationality (Euro-African, belonging to mixed families), while the parents of the remaining five are both African. One married another Sudanese and has a child. Two hold diplomas from institutions in Africa.

An interesting mark of integration concerns clothing; four of the young women dress in traditional attire, while five choose Western style garments. All, though at different levels, were capable of understanding and expressing themselves in Italian.

3. RESULTS

Socially, the students demonstrate a high level of integration; including those who, residing in the city for only a short period and continuing to experience difficulty and disorientation, are strongly optimistic about the future.

Four (out of five) were infibulated young East-Africans, knowledgeable and experienced regarding FGM, who respected the central theme of this project. None of the four girls with one Western parent were subjected to any form of FGM. This lack of direct experience, however, did not prevent their thorough knowledge and subsequent denunciation of the problem.

Among the infibulated subjects, one had undergone the operation twice: at age three, together with two elder sisters in her home, and alone at age five in a private clinic.

For the others, the intervention took place respectively at age six, age eight (alone in the home), and age ten (together with a cousin and a young aunt).

The infibulated girls have memories of both the intervention and the associated festivities traditional in the African context; all nine exclude the possibility of a future child undergoing excisory practices. In the interviews, breaching the subject of sexual experience was no simple matter; six women chose to avoid the question. Two openly declared never to have had sexual intercourse, and one would discuss the matter only in reference to the problems in relations with her husband, caused by the severity of infibulation undergone in a rural zone of Africa. This last case is the only woman to be married, with a child, and thus the only one to have faced birth and deinfibulation in Italy.

The question connected to the possibility of reversing the intervention solicited the highest degree of perplexity: three intact subjects were not familiar with even the meaning of the word or nature of the operation, while those infibulated (currently remaining infibulated) maintain that a possible deinfibulation would bring substantial acceptance problems with family and acquaintances, and that the choice to undergo such an intervention must always be made with the involvement of one's partner.

All of the young women consider the possible organization of conferences on FGM with enthusiasm. All are convinced of their efficacy in raising awareness of the issue and in representing valid alternatives to the arguments of the mass media, which is often less objective and characterized by an alarmist interest in the matter.

4. COMMENTS

The interviews underwent a purely qualitative analysis, as they were numerically insufficient to facilitate statistical elaboration. Nevertheless, they allowed us to reconstruct meaningful life elements. The subjects taken into consideration represent an elite group, as demonstrated by their high levels of education and their opportunity to be educated abroad.

Unusual and surprising for Africa, we came across a case of an African girl whose father suggested deinfibulation (not later completed), and a case of a young woman who, born to African parents, is genitally intact and lives in a family that has abandoned the mutilation tradition for generations.

Every story contains noteworthy elements, but the most significant result is that all infibulated students are born to two African parents, while those intact are of mixed heritage (in Afro-European families). This applies both for a Western mother or father.

Other key observations concern direct experience of infibulation:

- In accounts of the operation endured, principal figures performing the infibulation are always women (not men).
- The memory of the infibulatory intervention does not fade with the passage of time: two of the girls interviewed (particularly one who had undergone the experience twice) did not hesitate to admit that they could "still feel the scissors cutting."
- The memory of the pain remains tangible, a sensation implacable over time; only one girl denies having suffered. This tendency, common in Africa, may be interpreted by Westerners as emphasizing the social significance of the practice.

The terminology used to indicate FGM changes with the subject. Students who have been integrated into Italian society for a longer period and those of Western rather than African origin speak spontaneously of "infibulation," while young Africans tend to refer to the phenomenon with the word "circumcision," more commonly used on their native continent.

Regarding deinfibulation, only one of the four infibulated subjects is now deinfibulated and is happy to be so. She alone is married and chose to be deinfibulated with her partner's agreement after marriage, in preparation for the birth of her child (birth is one of the circumstances in which deinfibulation is recommended), and for reasons associated with the difficulty of sexual intercourse with her husband. The other three, without apparent physical repercussions connected to their condition, prefer to postpone a decision about such an intervention to a future time, with the agreement of their partners. Only a mutual decision stands a chance against possible stigmatization of their family.

Suggesting a complete and effective knowledge of the phenomenon, and of the girl's respective understanding, we found that all subjects denied that religion was the basis for FGM.

5. CONCLUSION

Reviewing the phases of our survey and the results obtained, we succeeded in creating an outline of the situation in which an African girl finds herself when placed in a society, such as Padua, that is decidedly different from her original social context. The specifics collected largely succeeded in mitigating limitations created by the small number of subjects.

Among the various elements, there emerged an insistence, which may serve as a guide for future initiatives, on the need to speak consistently of the FGM problem. Our subjects also recognized the importance of raising awareness of the issue in order to eradicate it. In this sense, conferences on this topic are considered to be effective. It is also vital to recognize the importance of gradual steps (alternatives to infibulation) toward the eventual elimination of these practices.

BIBLIOGRAPHY

Fusaschi M. I segni sul corpo. Per un'antropologia delle modificazioni dei genitali femminili. Bollati Boringhieri, Torino, 2003.

Grassivaro Gallo P. Figlie d'Africa mutilate indagini epidemiologiche sull'escissione in Italia. L'harmattan Italia ed., Torino, 1998.

Grassivaro Gallo P and Lineri S. Donna infibulata ed ostetrico. I protagonisti del parto in Italia. Unipress, Padova, 2003.

Lis A, De Zordo MR, Venuti P. Il colloquio come strumento psicologico. Giunti Ed, Firenze, 1995.

WHO, FRH, GWH Female Genital Mutilation. Integrating the Prevention and the Management of the Health Complications into the Curricula of Nursing and Midwifery. A Teacher's Guide. WHO, Geneva, 2001.

Chapter 8

FEMALE GENITAL MUTILATION AMONG AFRICAN IMMIGRANTS IN GREECE

The First Cognitive Study

Pia Grassivaro Gallo, Anna Iordanidou, Franco Viviani

Padua Working Group on Female Genital Mutilation, Department of General Psychology, University of Padua (Italy)

Abstract: This paper is the first cognitive survey documenting female genital modification in Greece. Seventy-three obstetrician/gynecologists completed a questionnaire and then were individually interviewed. The subjects work at Alexandra's Hospital in Athens and the General Hospital of Kavala (Macedonia) or were contacted at their private clinics. Results: 41% (n. = 30) of obstetrician/gynecologists had examined at least 95 adult mutilated patients for problems resulting from their mutilation. Few patients requested such operations. From an indirect estimation of the data supplied by the OAED (governmental organization for the concession of the working papers to immigrants), it would appear that 1,239 mutilated women reside in Greece, of which approximately one-half live in the capital. Among the problems faced during the specialist examination were difficulties in communication between doctors and the immigrant women and the surprise of the doctors when faced with excision. The Greek data are similar to Italian data obtained through a similar epidemiological survey in 1993-1994. Therefore, the problem of FGM in Greece is analogous to the same problem in Italy.

1. INTRODUCTION

Female genital modification (FGM) affects women in those non-Western countries where it is practiced. For many decades, Western countries have also been confronted with this problem due to the large waves of African immigrants who bring cultural practices such as FGM with them. Recent international surveys document the presence of FGM in ten European countries. Greece was excluded from these surveys (WHO, 1996: 28 and

G.C. Denniston et al. (eds.), Bodily Integrity and the Politics of Circumcision, 93—102.

Figure 1.1). Informal contact with representatives of nongovernmental organizations (NGOs) who work with immigrants (Institute for Equal Opportunities, Médicins Sans Frontières of Athens, Amnesty International Hellas) has permitted us to suspect that mutilated immigrant women have been in Greece since 1999. Even specific institutions in Athens, such as the Home Office, the Department of Citizenship; the Ministry of Foreign Affairs, the Department for Human Rights and Ethnic Groups, the Ministry of Health, and the Department for Maternity Rights had very little data related to immigrant African women (Table 4, column C). These institutions had no information about mutilated women. Research in the main libraries of the country (National Library of Athens, the Library of the Faculty of Medicine and Surgery, the Library of the Greek Parliament, and the National Institute of Statistics), together with specific on-line research in the various data banks of the National Institute of Statistics and the University of Thessalonica, has confirmed that no studies about FGM in Greece have been performed. The aim of this study is to correct this information deficit.

2. METHODS AND SUBJECTS

In order to obtain the required information, we administered an individual questionnaire to seventy-three obstetrician/gynecologists. This was followed by a personal interview. The questionnaire, based on one already used in Italy for similar surveys (Grassivaro Gallo, 1998), was translated and adapted.

In particular, sections A and B (with respectively eleven and eight items), collected information regarding the professional activity of the subjects. To complete the presentation, drawings of infibulation were made (Type 3 FGM) (Toubia & Izett, 1998).

When interviewing the subjects using the questionnaire, there was difficulty with the term *agtiriasmòs* (infibulation), which was completely unknown to many obstetrician/gynecologists. They assumed that the word was a mis-spelling of *acrotiriasmòs* (mutilation) since the words resemble each other. Paradoxically, it turned out that some of the doctors had examined infibulated women without knowing the meaning of the specific term.

Six private and state hospitals were contacted so that we could request interviews with their obstetrician/gynecologists, but only two answered positively: Alexandra's Hospital of Athens and the General Hospital of Kavala (Macedonia). The forty-three obstetrician/gynecologists contacted (58.8% of the total) are respectively 36.2% and 65.5% of those working within these institutions. The remaining 41.2% (thirty subjects) were collected with the "snowball" approach (Lis and Zennaro, 1997); the subjects were interviewed by appointment in their private clinics. Some of these doctors also worked in those state hospitals that had not given us permission to interview their staff (Elena's Hospital, Ira, IASO) (Table 8.1).

Table 8 1. Contacted hospitals/clinics and interviewed obstetrician/gynecologists

			Obstetrician/Gynecologists			
	Hospitals/ Clinics	Working in (N°)	Contacted		Hospitals/ Clinics	
			(N°)	%	Yes	No
	Elena (pu)	/*	3	/	2	1
	Alexandra (pu)	105	38	36.19	22	16
Attikh	Ira (pr)	400	5	1.25	0	5
(Athens)	Areteio (pu)	40	3	7.50	1**	2
	Iaso (pr)	/*	7	/	1**	6
	Private Clinics (pr)	/	10	/	4***	9
Macedonia (Kavala)	Gen. Hosp of Kavala (pu)	8	5	62.5	0	5
	Private Clinics (pr)	/	2	/	0	2

Total number of 73 subjects
pu = public hospitals/clinics
pr = private hospitals/clinics
* It was not possible to obtain official data
** Experience of mutilated women only in foreign countries (Italy, Libya)
*** Experience of mutilated women in Alexandra Hospital in their practice.

3. RESULTS

With reference to Table 8.2, among the interviewed subjects, there was a clear prevalence of male obstetrician/gynecologists (82.2%); their average age was 39.9 ± 7.5 years and they had been working for five to ten years in their field. The majority (67.1%) worked in a hospital setting, and it was there that they had been confronted with female genital mutilation. The obstetrician/gynecologists generally knew of the tradition; many (31.5%) had learned of it from university studies, and the majority (87.7%) hoped for laws to be passed against it.

Table 8.2. General characteristics of the obstetrician/gynecologists (N = 73). Age (in years): M ± D.S. = 39.86 ± 7.50. Professional activity (in years): M ± D.S. = 10.17 ± 7.44

Variables	Level of Variables	%
Gender	Male	82.2
	Female	17.8
Institute	Self employed	16.4
	Private clinic	15.1
	Hospital	63.0
	University (hosp/private)	5.5
Knowledge of FGM	Yes	98.6
	No	1.4
Knowledge of FGM practiced in Greece	Yes	4.1
	No	94.5
	NR(*)	1.4
Source of FGM knowledge(°)	High degree studies	31.5
	Publications	24.7
	Media	75.3
	Activity in developing countries	2.7
	Practice in Greece	26.0
	Activity in Western Countries	4.1
	Other	6.8
	NR(*)	1.4
Rationale for FGM(°)	Cultural	42.5
	Religious	41.1
	Men's control over women	17.8
	Sex desire reduction	11.0
	No emancipation	5.5
	Virtue protection	5.5
	Hygiene	4.1
	Ignorance	2.7
	Unknown	4.1
	NR(*)	1.4
Opinion on FGM regulation in Greece	To be legalized	1.4
	To outlaw	87.7
	To be discussed at governmental level	6.8
	Other	4.1
Visit to mutilated women	Yes	41.1
	No	57.5
	NR(*)	1.4

(*) NR: No Reply

(°) Several replies obtained

Of the doctors interviewed, 41.1% answered that they had examined mutilated women (of which only 23.3% were infibulated). These thirty obstetrician/gynecologists (Table 8.3) worked in the capital and had treated not less than ninety-five women suffering from gynecological complications as a result of mutilation. No mutilated children had been examined. A very small percentage of patients requested doctors to carry out mutilation operations: four requests for deinfibulation and two for postpartum reinfibulation, all of which were granted; one request for excision and one for infibulation, which were not granted. The questionnaire also included some items regarding hypothetical requests for excision operations from immigrant women: all the obstetrician/gynecologists declared that they would refuse to carry them out, while they would be prepared to carry out deinfibulation. Finally, it was pointed out that, during the specialist examination, a remarkable variety of problems emerged, such as communication difficulties due either to language or to a lack of technical (gynecological) knowledge. Contextually, the behavior of the patients varied a great deal, some displaying fear, agitation, shame, and exaggerated pain.

Table 8.3. Obstetrician/gynecologists (N =30) faced with mutilated women. Examined women: age (in years) M ± DS = 30.15 ± 4.76

Variables	Level of Variables	%
Number of examined women	1- 5	80.0
	6-10	13.4
	NR	6.6
Types of FGM(°)	Infibulation	23.3
	Clitoridectomy and/or Excision	83.4
	Other	23.3
FGM complications	Yes	83.3
	No	10.0
	NR(*)	6.7
Types of complications(°)	Gynecological	60.0
	Sex-psychological	6.7
	Gynecological-urinary	3.3
	Other	13.3
	NR(*)	16.7
Problems that emerged during the check-up	Communication difficulty	30.0
	Surprise and/or detest	26.7
	Difficulty in dealing with the situation	3.3

Variables	Level of Variables	%
	All the types	6.7
	None	6.7
	NR(*)	26.6
Patient's behavior during the check-up	Fear and/or agitation	46.7
	Exaggerated manifestation of pain	6.7
	Shame	6.7
	Fear+lack of trust	3.3
	Fear+agit.+exaggerated pain	6.7
	Spontaneous	6.7
	NR(*)	23.2

(*)NR: No Reply

(°): Several replies obtained

4. A COMPARISON BETWEEN GREEK AND ITALIAN RESULTS

We have already specified how the study in Greece followed the lines of a similar study carried out in Italy in 1993-1994 (Grassivaro Gallo, 1998). Being aware of the differences between the situations in which the studies were carried out, we would like to offer a comparison.

Subjects: the Italian obstetricians came from all over Italy; the Greek ones came from Athens only, where 44.3% of the total number of immigrants is concentrated (164,000 immigrants out of a total of 369,629, of which 93,482 were women – EIE National Institute of the Work) (Kokkaliari, 1999). Note that figure of 369,629 corresponds to the number of immigrants who had requested work permits. In the study carried out in the region of Macedonia in Kavala, no cases of FGM were found (Table 1).

4.1 Presence and Type of FGM

In the two Mediterranean countries, the practice of genital excision exists among African immigrants. While evidence of this was also found in Italy, through the direct experience of doctors (hospital casualty units), in Greece, only indirect information emerged. In both countries, gynecologists of African origin and/or mixed culture were indicated as having carried out different types of mutilation. The medical profession unanimously demands a legal regulation that in Italy just has been put into effect through the propagation of guidelines (Dipartimento per le Pari Opportunità, 2001).

Altogether, one out of two doctors in both countries had come into contact with mutilated women: 46% of Italians and 41% of Greeks (Table 2). The Italian doctors, however, seemed to have more experience; many of them had seen from one to ten cases and some up to one hundred. The Greek obstetrician/gynecologists had seen very few cases. Moreover, the Greek doctors had never seen cases of child or adolescent mutilation, whereas this has been documented in Italy since 1990. The African women in the two countries present various types of FGM since their countries of origin are different. In Greece, infibulation is rarer, and therefore the physical consequences are less serious.

Both surveys confirm the discrepancy that exists between the presumed number of mutilated women in the country and the scarcity of requests for mutilating operations. This, therefore, confirms the suspicion that mutilating operations occur clandestinely among the African diaspora. Doctors, however, generally agree that they are prepared to operate in cases of deinfibulation only.

4.2 Doctors Confronted with FGM

Among the problems that emerged during the specialist examinations with mutilated women were communication difficulties. Doctors were also shocked when faced with excision for the first time. At times, this shock caused them to make poor medical decisions.

5. COMMENTS AND CONCLUSION

We would like to emphasize that the information presented is new. This is the first study that documents the existence of FGM in Greece. We are aware of the limitations of our methodology, therefore, this should be considered a pilot study. Further studies of the problem will be conducted in the future.

Of the total seventy-three subjects, the sixty-six interviewed in Athens correspond to 6.1% of the total number of active professionals in the city (1,065 units according to the Secretary of Obstetricians, 2000), where the majority of immigrants live (Kokkaliari, 1999). Alexandra Hospital, the only hospital in the capital to give permission to conduct our interviews, is one of the hospitals better equipped to receive immigrant African women residing in Athens. An attempt to include the General Hospital of Kavala (Macedonia) was unsuccessful. We conclude that the data pertains only to Athens but accurately reflects the entire country because most African immigrants live in Athens.

The second part of the study concerns the existence of FGM in Greece as confirmed by forty-one percent of the obstetrician/gynecologists who had treated no less than ninety-five mutilated patients. These figures, however, are insufficient to determine the exact number of mutilated African women in Greece. Therefore, we made an indirect estimate using data supplied by the OAED (governmental organization for the concession of working papers) (Table 8.4).

Table 8.4. Estimated number of mutilated African women in Greece in 1996, distributed according to country of origin

A	B	C	D
African Mutilating Countries Present in Greece(1)	% of Mutilated Women (1)	Immigrant Women (2)	Estimated Mutilated Women
Egypt	97	347	337
Nigeria	40	350	140
Ethiopia	90	634	571
Ghana	30	77	23
Sudan	89	25	22
Tanzania	10	28	3
Kenya	50	114	57
Sierra Leone	90	95	86
Total	/	**1,670**	**1,239**

(1) according: WHO, 1998
(2) with work permit in 1996 (OAED)

Table 4 contains the following information: in columns A and B, the percentages of FGM for every African country represented in Greece (according to the most recent bibliography available, Toubia and Izett, 1998); these vary from a minimum of ten percent for Tanzania to a maximum of ninety-seven percent for Egypt. Column C contains OAED data. Multiplying the values in C by the percentages in B, we obtained (in column D) the estimated number of mutilated women in Greece, a figure of 1,239 out of 1,670 immigrants from the countries that practice excision. Moreover, if the ninety-five mutilated women examined by the obstetrician/gynecologists of Athens were still in Greece, it can be estimated that one out of thirteen mutilated women should have made contact with a specialist, equal to 7.66% of the total, a percentage similar to that calculated for Italy (7%, Grassivaro Gallo, 1998). The above estimate of 1,239 mutilated women constitutes the minimum number for the country since it excludes immigrant women legally residing in Greece without a work permit who, therefore, are untracked by the OAED. This number also excludes the large number of illegal immigrants. Given these estimates and the small number of requests for FGM made to obstetricians, we conclude that the practice of FGM in Greece is substantially clandestine.

We have already noted that Greek obstetricians did not report seeing any mutilated girls, whereas, in Italy, various cases of this kind have been reported since the early 1990s in hospital casualty units (Grassivaro Gallo, 1998). This difference may be due to the recent history of immigration to Greece, as well as to the different types of FGM practiced as a result of the various ethnic origins of the immigrants, and consequently the greater difficulty for the doctors to observe excision, which is not as obvious as infibulation.

Considering the present evolution of the phenomenon in Italy, especially regarding those children who are at risk of FGM, we can assume that similar situations will occur in Greece in the near future. Examples of these situations are the importation of African obstetricians to carry out operations on children in Greece; the establishment of clandestine and illegal clinics for the practice of FGM in an uncontrolled way (in the interviews collected in Athens, there was mention of an Egyptian gynecologist who practices FGM clandestinely); the presence of mixed-race couples (Greek mother/African father), where the traditions of the father are followed and the children are sent to relatives in Africa to be mutilated; the filing of lawsuits over FGM.

In general, Greek obstetricians are acquainted with FGM but lack knowledge about its cultural dimensions. This leads them to adopt an attitude of non-responsibility toward immigrant women, considering the practice of excision as "their own affair," "their traditions," etc. Also, the numerous problems faced when examining mutilated women testify to a lack of information. In fact, 66.7% of the obstetrician/gynecologists who had examined mutilated women reported having encountered difficulties during their contact with them. The most frequent difficulties were communication (30%), shock, and/or disgust (26.7%). Moreover, considering the behavior of the patients during the examination, as reported by the doctors (46.7% reported reactions of fear and/or agitation), it could be assumed that the inner conflict experienced by the obstetrician when faced with FGM has a negative influence on the reactions of the women.

It would, therefore, be important to instigate an education program for those doctors who are more likely to come into contact with FGM. Such a course would include information about anthropology, cross-cultural psychology, the ethical-deontological aspects of FGM, and about the medical treatment of problems caused by genital mutilation.

From a comparison of the Italian and Greek data, only a few differences emerged compared to the considerable similarities. We, therefore, can assume that, in Greece, the phenomenon will follow the same line of development as in Italy, although more rapidly. In fact, the Greek female obstetrician/gynecologists interviewed in the year 2000 were more inclined to discuss and to inquire about FGM compared to the Italian female obstetrician/gynecologists interviewed between 1993 and 1994 (Grassivaro Gallo & Cortesi 1999). Moreover, while the Italian doctors had largely become aware of the existence of female genital mutilation through the shock of treating mutilated African patients, the Greek obstetricians had already learned of the problem through the mass media. Therefore, we suspect that the presence of mutilated women will not be as surprising to Greek doctors. They therefore will be better prepared than Italian obstetrician/gynecologists.

In conclusion, the Italian experience with FGM may help Greece to act preemptively to establish professional guidelines to deal with the problem of FGM.

We are grateful to all those who have helped us conduct this research project: Professor S.P. Michalas, Director of the Department of Obstetrics and Gynecology of the Alexandra Hospital in Athens; N. Georgacopoulou and M. Kaliudaki, heads of NGOs in Athens; G. Mitragas, Associate Professor of Bibliothiconomy of the University of Thessalonika

BIBLIOGRAPHY

Dipartimento per le Pari Opportunità. Presidenza del Consiglio dei Ministri. Mutilazioni Genitali Femminili: cura e prevenzione. Linee guida per operatrici e operatori sanitari, sociali ed educativi. Istituto Poligrafico e Zecca dello Stato, Roma, 2001.

Grassivaro Gallo P. Figlie d'Africa mutilate. Indagini epidemiologiche sull'escissione in Italia. L'Harmattan-Italia, Torino, 1998.

Grassivaro Gallo P. and Cortesi C. Linee guida per il personale medico di fronte a casi di Mutilationi Genitali Femminili (FGM). Messa a punto delle iniziative a livello nazionale ed internazionale. Osservatorio Regionale Immigrazione Veneto (ORIV). Quaderni di Ricerca, 5:1-55, 1999.

Kokkaliari K. Polietniki. Employment Market Ethnos, 7 Decembers 1999.

Lis A, Zennaro A. Metodologia della ricerca in Psicologia Clinica. La Nuova Italia Scientifica, 1997.

Toubia N, Izett S. Female Genital Mutilation: an Overview. World Health Organization, 1998, (web site: http://www.who.int/dsa/cat98/fgmbook.htm, 17 April 2001).

WHO Female Genital Mutilation. Report of WHO Technical Working Group, 17-19 July 1995. Geneva, 1996.

Chapter 9

DEINFIBULATION IN ITALY

Pia Grassivaro Gallo, Lara Franco, Lisa Rivaroli
Padua Working Group on Female Genital Mutilation, Department of General Psychology, University of Padua (Italy)

Abstract: The objective of this investigation is to construct a detailed and current map of institutions and medical professionals in Italy that perform deinfibulation, thereby creating a useful resource for East-African immigrants in Italy. Two types of questionnaires were used for interviews of 309 obstetricians and gynecologists. Information collected included the level of familiarity and experience that health professionals have with the intervention, as well as their opinions on its utility and optimal timing. Of the 309 obstetricians and gynecologists we identified: 41 deinfibulators (subgroup A), from which we extract 9 non-occasional deinfibulators (subgroup B), who performed more than 10 deinfibulations. Characteristics of subgroup B: subjects, female or male, are mature in age and profession; think that deinfibulation will improve sexual health and life and will resolve pathological genital and urinary sequelae; are knowledgeably about female genital mutilation because of his/her origins or professional experience in developing countries in East Africa; and are able to establish a good rapport with patients, enabling them to provide information about deinfibulation. In conclusion, the practice of deinfibulation in Italy was found to be rare (10% of professionals interviewed), largely because it is not proposed by doctors or requested by women.

INTRODUCTION

The following research was inspired by a recent paper about obstetrical problems among infibulated women that was published by the Padua Working Group on Female Genital Mutilation (FGM) (Grassivaro Gallo and Lineri, 2003).

G.C. Denniston et al. (eds.), Bodily Integrity and the Politics of Circumcision, 103—115.

From this study, we found that obstetricians and gynecologists consider deinfibulation to be of critical importance for enabling infibulated women to avoid many of the physical sequelae of infibulation, and for improving their sexual life, although the primary importance of this intervention seems to be to make it possible for infibulated women to give birth. Deinfibulation is also considered very important by the European (WHO, 2001) and Italian guidelines (Gori *et al.*, 2003) for the managment of pregnancy, childbirth, and the postpartum period in the presence of female genital mutilation.

1. SUBJECTS AND METHODOLOGY

Between 2000 and 2004, we administered a general questionnaire to 309 obstetricians and gynecologists who were contacted at various medical conferences, principally in Emilia Romagna, the Veneto, Lombardy, Friuli, Venice, Giulia, and Tuscany; the general questionnaires enabled us to extend the research to obstetricians and gynecologists who are living in other regions, for a total of 309 professionals.

The objective of the general questionnaire was to understand how Italian professionals deal with excised women, whose numbers are growing in Italy due to the increase in immigration from East Africa. Our main objective was to find the proportion of doctors who have a thorough understanding of deinfibulation and who practice it. A detailed analysis of results has been presented in a project that is now running in the "Premio G. Gherson 2004" competition (Catania, *et al.*, unpublished data).

Of those interviewed, forty-one obstetrician/gynecologists (13.3%) performed at least one deinfibulation: Subgroup A. To this subgroup, we administered in-depth interviews (specific questionnaires) and discovered that deinfibulation is practiced in Lombardy, the Veneto, Trent, Alto Adige, Emilia Romagna, Lazio, Tuscany, and Campania. Within Subgroup A, we found nine professionals whom we call "non-occasional deinfibulators" (Subgroup B), who performed ten deinfibulations or more. We only analyzed the responses obtained from Subgroup B in order to indicate the characteristics of the professionals who treat and who have gained the trust of East-African women.

The content analysis methodology we used seemed to be the most appropriate to codify the information obtained from interviews. This technique allows us to analyze texts (the written reports of interviews) through a chart that focuses the single units in relation with variables obtained from the content categorization. After the creation of categories (and eventual subcategories) on the basis of what the subjects said, every document is decodified in order to create a chart that can be read in statistical terms (Denzin and Lincoln, 1994).

2. RESULTS

Subgroup A: Forty-one obstetricians and gynecologists who had at least one experience with deinfibulation; general characteristics are summarized in Table 1.

Table1. Subgroup A (N = 41)

Variables	Modality	N°
Sex	Female	20
	Male	21
Nationality	Italian	36
	East-African	4
	Others	1
Professional Degree	Obstetrician	6
	Gynecologist	28
	General Doctor	7
Clinic Types	Asl (Public hospital)	38
	Others	3
Sources on Deinfibulation (°)	Published articles/mass media	11
	Conferences	7
	Colleagues	12
	Italian experiences	20
	East-African experiences	12
	Other/specific seminaries	10
	NA	5

Average age (years): 50.2+9.4

Average age (years) of professional activity: 22.2+9.

NA = No Answer

(°) = Multiple answers were allowed

The median age is 50.2 (±9.4) years; years of career: 22.2 (±9.6); equally males and females (respectively, twenty-one and twenty). Of these, there were twenty-eight gynecologists (ten women), six obstetricians (five women), and seven general practitioners. Thirty-eight subjects work in public hospitals, one in an outpatient office, and two in private hospitals. Most of them are Italian, with the exception of three Somalis, one Eritrean, and one Chilean.

All forty-one subjects had treated excised females; thirty-nine subjects reported examinations in public clinics (*Asl*); the number of women examined varies between 0 and 200 (standard deviation). The standard deviation of ± 48.8 shows the difference between professionals who had sporadic experiences and those favored by immigrant communities. Among those who had examined more than one-hundred excised women, we found two Somalis, one Italian-Eritrean, and two Italians. Only seven reported examination of children. The Somalian and the Eritrean mentioned a total of twenty girls. All the others reported less than ten. The types of mutilation seen in those examinations were primarily infibulation (seen by thirty-six professionals) and total clitoridectomy (observed by fourteen professionals); clitoridectomy with labia minora excision (observed by thirteen professionals), and excision of the clitoral hood (observed by eleven professionals). Information on specific experiences of the deinfibulated are presented in Table 2.

Table 2. Obstetrician/Gynecologists who performed at least one deinfibulation: Their Experiences. Subgroup A (N = 41)

Variable	**Modality**	**N**
Deinfibulation: quantity	One	13
	Some	17
	Many	9
	NA	2
Deinfibulation technique	Traditional surgery	37
	Laser	3
	Others	1
Requests for Deinf.		25
	No	16
	NA	4
Where Deinf. was performed (°)	Asl (public Hospital)	38
	Private Clinic	2
	Private Outpatient Clinic	1
	NA	1
Deinf. motivation (°)	Delivery	29
	Gynecologic diseases	12
	Sexual diseases	12
	Pregnancy	8
	NA	4

Variable	Modality	N
Who suggested Deinf. (°)	Patient	7
	Doctor	20
	Together	17
	NA	6
Patient's reaction (°)	Immediate acceptance	19
	Refusal	3
	Initial confusion	7
	Need for family's/partner's agreement	7
	NA	12
Was Deinf. explained?	Yes	22
		7
	NA	12

As for the number of deinfibulations performed, we separated professionals according to the experiences they had with deinfibulation and according to the responses provided: one/some/many (ten or more deinfibulations).

Thirteen subjects (30.9%) had only one experience, while nine (19%) reported many. We called the latter "non-occasional deinfibulators."

2.1 Deinfibulation Techniques

Standard surgical technique was reported by thirty-seven subjects; three reported laser surgery; only one professional reported using Loop electrosurgical excision procedure (Leep). The health facilities where deinfibulation was known and practiced are: Public Clinics (thirty-eight), one Private Clinic in Pisa, another near Naples, and an outpatient clinic in Verona.

2.1.1 Motivations that led to the choice of deinfibulation (multiple answers were allowed)

With decreasing frequencies, the following answers were obtained: delivery, gynecological diseases, sexual diseases, pregnancy, respectively from twenty-nine, twelve, twelve, and eight subjects.

The above-mentioned responses confirm that delivery is often the only time doctors encounter excised women; consequently, deinfibulation is necessary to make delivery possible and is not performed for other medical reasons, such as the pathological consequences of FGM.

2.1.2 Who suggested deinfibulation, the physician, the woman, or the two together?

The decision to perform deinfibulation was made by both doctor and patient in seventeen cases. In twenty cases, it was suggested by the physician. In seven cases, it was requested by the patient. Six professionals did not respond to the question, perhaps because they did not have any time to discuss deinfibulation with the woman due to the urgency of the situation.

2.1.3 Patient's reaction during the discussion and recommendation for deinfibulation (multiple answers were allowed)

In nineteen cases, there was immediate acceptance (45.2%). Seven women needed to discuss it first with their partners and/or families (19%). Seven women were initially confused (16.7%). Three flatly refused (7.1%).

In order to best serve East-African women, we compiled a preliminary list of Italian hospitals and clinics where deinfibulation is performed. The list is admittedly incomplete. Names in boldface type indicate facilities where "non-occasional deinfibulators" work:

2.1.3.1 Lombardy

Milan: S. Paolo Hospital: Div. Ostetrica

Mangiagalli Hospital: Div. Ostetrica

NAGA (ONG) Clinic: Div. Ostetrica

Mantova: Carlo Poma Hospital: Clinica Ostetrica

Brescia: Spedali Civile Hospital: Clinica Ostetrica

2.1.3.2 Piedmont

Torino, we have no references about the clinics

Mondovì (Cn), we have no references about the clinics

2.1.3.3 The Veneto

Padua: Ospedale Civile Hospital: Divisione Ostetrica; Policlinico Universitario: Clinica Ostetrica

Verona: Borgo Roma Hospital: Clinica Ostetrica

Bovolone (VR) Civil Hospital: Reparto di Ostetricia

2.1.3.4 Friuli-Venezia-Giulia

Udine: Policlinico Universitario

Trieste: Civil Hospital: Clinica Ostetrica

2.1.3.5 Emilia Romagna

Bologna: Maggiore Hospital: Clinica Ostetrica

S. Orsola Hospital: Clinica Ostetrica

Ferrara: S. Anna Hospital: Clinica Ostetrica

Forlì: Pierantoni Hospital: Clinica Ostetrica

Modena: Civil Hospital: Clinica Ostetrica

Sassuolo (Mo): Civil Hospital: Clinica Ostetrica

Parma: Policlinico Universitario: Clinica Ostetrica

2.1.3.6 Tuscany

Careggi (Florence): Policlinico Universitario: Regional Centre for MGF

Siena: Azienda Ospedaliera Universitaria Senese: Clinica Ostetrica

2.1.3.7 Umbria

Terni: S. Maria di Terni Hospital: Divisione Ostetrica

2.1.3.8 Lazio

Rome: Istituto San Gallicano (IRCCS), Umberto 1° Hospital: Clinica Ostetrica

S. Camillo Hospital: Clinica Ostetrica

Grassi Hospital: Clinica Ostetrica

Policlinico Gemelli: Clinica Ostetrica

2.1.3.9 Campania

Naples: Policlinico Universitario Federico 2°: Clinica Ostetrica

Private Clinic 'Mediterranea'

Caserta: Annunziata Hospital

Subgroup B: Nine obstetricians and gynecologists performed ten or more deinfibulations. We designate them as: "non-occasional deinfibulators."

Subgroup B is composed of five women (two obstetricians: one is Somali and works in Parma, one is Italian and involved in voluntary services for immigrants in Bologna; one gynecologist is Somali and works in Rome; two are Italians, involved in voluntary services and political groups supporting immigrants who work in Milan and Rome) and four males (one Somali is director of the Regional Centre for FGM in Careggi, Florence; one is Chilean, employed in outpatient clinics in Florence; two are Italian, one working in Rome, the other in Verona.

2.1.4 In-depth interviews analysis (Table 3)

As explained above, every professional had treated more than one excised woman. This means that single sub-categories are not mutually exclusive. We assigned the value one for each experience identified into a category reported by the professional.

Table 3. Non-Occasional Deinfibulators. Information emerged from in-depth interviews Subgroup B B (N = 9)

Modalities	**Yes**	**No**	**NA**
DEINFIBULATION			
Deinfibulation requested for sexual diseases	9	-	-
Deinfibulation requested for gynecological disease	9	-	-
Deinfibulation requested before childbirth	9	-	-
Discussions over deinfibulation possibility	9	-	-
Surgically mutilated girls seen in Italy	4	-	5
WOMEN'S BEHAVIOR			
Woman was against deinfibulation	6	2	1
Woman's partner was against deinfibulation	4	1	4
Request for *post partum* re-infibulation	4	5	-
Women were sent by colleagues	8	-	1
Women arrived without apparent contact	4	4	1
Women arrived on other women's suggestion	8	-	1
OPINIONS on the INTERVENTION			
Deinf. may reduce pain during sexual intercourse	9	-	-
Deinf. may reduce genital and urinary complications	9	-	-
Deinf. is useful as soon as possible for every excised woman	9	-	-
Deinf. should be done during the first trimester of pregnancy	7	-	2
Deinf. should be done during childbirth	5	2	2
Deinf. is necessary for a successful delivery	3	3	3
Italy needs a law to condemn circumcisers	8	-	1
Some colleagues will circumcise if paid	5	-	4
SOURCES ON DEINFIBULATION			
Direct experience and personal interest	9	-	-
From colleague to colleague	7	1	1
Specific seminars	6	2	1
Experience as health workers in East Africa	6	3	-

Subjects average age C: 47.6+12.9

Average length of professional activities (years) :20.6+12.14

Our categories are:

2.1.4.1 Deinfibulation: knowledge and experiences with the intervention

All nine subjects reported that the women had requested deinfibulation to cure gynecological and sexual pathologies or to improve their chances of giving birth successfully. Many responses, however, indicate that the majority of excised women still enter labor infibulated.

All nine subjects report having discussions with their infibulated patients in order to introduce them to the possibility of deinfibulation.

Four reported seeing cases of infibulation that had been performed by a surgeon (we don't know if surgical infibulation was performed in Africa or in Italy).

2.1.4.2 Patient's behavior during visits and counseling

Six professionals reported encountering women who objected to deinfibulation. Four refusals came from the woman's partner.

2.1.4.3 Professionals' opinions on the utility of deinfibulations

All nine subjects believe that deinfibulation improves sexual life, decreases pain during intercourse, reduces pathological genito-urinary sequelae of infibulation. They also recommend that infibulated women undergo deinfibulation as soon as possible. Six believe that the first trimester of pregnancy is the ideal time for deinfibulation. Three, all Italian, claim that the intervention is unnecessary before childbirth and should be performed during labor. Five performed deinfibulation only during delivery. The others were concerned that deinfibulation during childbirth could harm the child.

In interpreting these answers, one must note that East-African women living in Italy rarely seek medical care during pregnancy. Consequently, there is seldom time to perform a preventive deinfibulation. In any case, only three (Italians) affirmed that deinfibulation is absolutely necessary in order to have a delivery free of complications for both mother and child. The others argue that deinfibulation is useful but not necessary because, in their opinion, the extended labor of excised women is rarely a problem.

Eight health workers believe it is necessary for the Italian government to pass a law criminalizing female circumcisers. They also argue for extensive outreach to East-African communities in Italy in order to educate, elevate, and emancipate them. Once this foundation had been laid, eradication can begin. Five doctors have no doubt about the presence in Italy of colleagues who would perform illegal circumcision if paid enough money.

2.2 Relations Between Health Workers and Circumcised Women

Eight subjects reported that women themselves are the gateway to reach a women's community. Sometimes the connection is through other doctors, who send the women to the most experienced colleague. Only four report occasional cases due to the presence of East-African communities in town. One (from Verona) asserts that all the women he attended in his clinic were there for reasons unrelated to infibulation. In these cases, we valued the answer as zero (no answer).

2.3 Sources for Learning About Deinfibulation and Training

All subjects learned 'in the field' (six directly in East-Africa, working in local hospitals) or personally documented the management required for circumcised women. Seven report having collected information from other colleagues. Six attended medical courses about FGM. Health workers also need to adopt a transcultural approach in order to communicate effectively about the tradition of female genital modification. It is interesting that the need for thorough knowledge is expressed by professionals that we consider to be "non-occasional deinfibulators."

3. CONCLUSIONS

This study about the ways that Italian healthcare facilities handle genitally mutilated women is incomplete because information has not yet been collected from all regions in southern Italy or from the principal Italian islands, i.e., Sicily, Sardenia, etc.

Our objective was to compile a list of facilities to be helpful to East African women in need of gynecologic and obstetric assistance. The list above indicates the regions covered by our research and the hospitals where deinfibulation is performed by "non-occasional deinfibulators" in Milan, Bologna, Parma, Florence, and Rome.

We also found that requests for deinfibulation were uncommon. The reason for this may be the absence of discussions between doctor and infibulated patient because the woman is unable to speak Italian or English. Other reasons may be ignorance on the part of the healthcare worker about FGM and the need for infibulated women first to seek approval from husbands and family. Therefore, deinfibulation is rare and is performed primarily in emergency situations.

We found only one healthcare facility where deinfibulation is routinely offered: the Regional Center for Curing and Preventing FGM Complications in Careggi, Florence, directed by Dr. Omar Hussein Abdulcadir (Somali) and his wife, Dr. L. Catania. In the past two years, more than sixty deinfibulations have been performed in this center. Requests for the intervention are increasing. This seems to be the only Italian facility providing services tailored to the needs of circumcised women.

The goal of the center is to offer healthcare to women, to prevent FGM sequelae, to develop a transcultural dialogue to enable immigrant communities to understand that they do not need bloody and cruel rituals to maintain their identity, to organize meetings that provide information about the subject, to promote specific training courses that will enable health workers to treat excised women, and to cooperate with other facilities that work for the eradication of female circumcision.

The center often treats immigrant women who are in the country illegally. During the first visit, the doctor attempts to establish a good rapport with the patient by inquiring generally about the patient's physical and psychological health. The doctor will initially avoid discussing female genital modification. The provision of healthcare is organized with a specific protocol: general analysis, obstetric and gynecologic examination, classification of the mutilation, and discussion on deinfibulation.

Obstetric deinfibulation is performed during delivery to open a passage for the baby. This intervention requires incising the infibulation scar and lateral episiotomy if necessary. Gynecological deinfibulation is performed to cure urogenital sequelae or sexual diseases. Laser surgery is the most appropriate technique for this type of operation.

The center in Careggi is probably based on London's Central Middlessex Clinic's facility, where English gynecologist P. H. Gordon worked. From 1993 to 1996, Gordon performed 107 deinfibulations. His medical staff included a Somali translator and a Muslim gynecologist (Zanuttini, 1997; Grassivaro Gallo, 1998).

Our questionnaire was filled out by 309 health workers, but we found only nine who performed more than ten deinfibulations. The "non-occasional deinfibulators" work in large city hospitals and often become a referral source for colleagues and women living in the area.

Most "non-occasional deinfibulators" have been working for more than twenty years. They believe that deinfibulation improves sexual life and reduces genital and urinary sequelae. They recommended it for all women who suffer from complications of female genital modification and will perform deinfibulation during childbirth if necessary (Table 3).

All of the "non-occasional deinfibulators" are thoroughly acquainted with FGM; three because it is part of their own culture, six because they had experience in the field, working in local East-African hospitals.

The "non-occasional deinfibulator" can establish effective communication with patients, the first step of which is to educate and later discuss female genital modification. Information imparted in this manner often reaches other women in the community and sometimes their husbands.

There is an obscure point, also mentioned in previous work (Grassivaro Gallo, 1998). Our subjects reported seeing infibulations that looked as if a surgeon had performed them. These infibulations had not been performed with traditional instruments. The doctors do not know whether these surgical infibulations were performed in Africa or in Italy. Moreover, it is very disquieting that five out of nine subjects suspect that there are doctors who would perform circumcision if paid enough money.

BIBLIOGRAPHY

Catania L., Omar Abdulcadir Hussein, Funghi P, Grassivaro Gallo P, Franco L, Rivaroli L, Simone S. La deinfibulazione in Italia, 2004 (unpublished data).

Commissione Nazionale Contro Le MGF, Dipartimento Per Le Pari Opportunita' (a cura di). MGF, cura e prevenzione. Linee guida per operatrici e operatori sanitari, sociali ed educativi. Presidenza del Consiglio dei Ministri. Dipartimento per le Pari Opportunità, Roma, 2001.

Denzin NK and Lincoln YS. Handbook of Qualitative Research. Sage, 1994.

Gori *et al* Le MGF nella popolazione immigrata. Raccomandazioni per i professionisti. Regione Emilia Romagna, 2003 (www.saperidoc.it/doc_120.html).

Grassivaro Gallo P, Figlie d'Africa mutilate. Indagini epidemiologiche sulla escissione in Italia. L'Harmattan-Italia, Torino, 1998.

Grassivaro Gallo P and Lineri S. Donna infibulata ed ostetrico. I protagonisti del parto in Italia. Unipress, Padova, 2003.

WHO. Managment of pregnancy, childbirth and the postpartum period in the presence of female genital mutilation. Report of a WHO Technical Consultation. Geneva, 2001.

Zanuttini P. Dossier: La ferita originaria. *Il Venerdi di Repubblica*, marzo/aprile, pp. 77-88, 1998.

Chapter 10

RESEARCH CENTER FOR PREVENTING AND CURING FGM AND ITS COMPLICATIONS

Abdulcadir Omar Hussen
Department of Gynecology and Obstetrics, University of Florence (Italy)

1. INTRODUCTION

At the Department of Gynecology and Obstetrics at the University of Florence, all the types of female genital mutilation (FGM) have been treated every day for more than twenty years; from the most severe and painful to the less severe ones. We receive hundreds of mutilated women, coming from all over the world and from local communities to have their health problems treated. Because of this, at the end of 2003, the regional authority recognized this work, and our department was named "Research Center for Preventing and Curing FGM and its Complications."

2. OBJECTIVES OF THE CENTER

2.1 Meetings

The center has regular meetings with the Association of Immigrants and their representatives from African countries and regions where FGM is still practiced. The first and foremost objective of these meetings is to prevent FGM by establishing relationships with influential religious and medical figures within these communities. The support and backing that these influential figures give the Research Center is fundamental in facilitating the

G.C. Denniston et al. (eds.), Bodily Integrity and the Politics of Circumcision, 117—122.

center's work of informing immigrant communities of the grave dangers of FGM. Special attention is given to families with daughters born around 1995 who will now be at the critical age for a mutilation ceremony. The summer holiday period is a key time, as families take their daughters back to their homeland or to other countries tolerant of this practice.

In our work with the immigrant community, the Center stresses the futility of existing ritual practices and how important complete undamaged genitalia are for a woman's health and for her ability to have children successfully. On this subject, our work with the men in these communities is very important. They must be taught what intact female genitalia look like and what is natural physical virginity, as many men believe that only mutilated women are virgins and refuse to marry non-mutilated women.

Another important issue to explain to people in these communities is the different way this practice is viewed in their host country. Immigrants need to know that this practice is illegal in their host nation because it is seen as a violation of rights, as violence against women, and as child abuse. The necessity to abide by the laws of the host nation is explained and emphasized.

All of this information is circulated through the preparation and distribution of information leaflets, especially during antenatal care. Our research has found that many women, as a result of the pain and suffering they have endured as victims of FGM, are searching for an alternative ritual for their daughters, something that does not mutilate but retains the symbolic ritual of the established practice.

2.2 Census

A census has been compiled of all the female babies born at the hospital of Careggi, Florence, over the past few years with the aim of preventing FGM. We hope that this project will be developed soon to include other hospitals in Tuscany.

2.3 Treatment for FGM complication

Complications due to FGM in women are treated, and material is being prepared regarding such complications as a guide for other doctors, obstetricians, and students of medicine.

2.4 Support

The Center provides a support and consultancy role for immigrants and their gynecologists in the Florence area. The Center is often approached for help and advice from doctors who have patients with problems beyond their experience or who need help with language or knowledge of equipment for treating FGM patients. As a result of this, the Center is going to prepare seminars and training courses.

2.5 Consultations

The Center often receives telephone calls from Italian doctors who need help and advice on FGM cases.

2.6 Treatment and Training

The Center treats and cures women with FGM during antenatal care. At the same time, we:

- Work on the prevention of FGM, included in antenatal care.

- Provide couples with information about the right time and form of any operation to reverse FGM (deinfibulation) and any risks that this may incur.

- Build a trusting relationship between the couple and the doctor before any physical examination or operation is performed.

- Diagnose the type and extent of the mutilation at the first visit.

- Establish guidelines for any operation to open type 3 mutilations.

- Train doctors and obstetricians in the different surgical techniques to reverse different types of FGM.

- Train doctors and obstetricians in ways of counseling women and couples with FGM.

2.7 Labor Support

The Center provides support for the laboring woman.

- Strategies for labor have to be developed, including ways of performing episiotomies or carrying out deinfibulation on women to reduce any trauma to the perineum during labor.

- Train staff in the labor room to recognize different types of FGM, making them aware of the complications involved if they have not been resolved during antenatal cure.

- Guidelines for the suturing of the perineal area in women with FGM during labor need to be established.

2.8 Postpartum Care

The Center provides post delivery care. Many FGM women need extra check ups after delivery to ensure a quick diagnosis of any ensuing complications.

2.9 Education

The Center has an active teaching program. Our seminars include:

- Discussion about our knowledge and the results of our research concerning all the areas of FGM. These seminars have already been held in many places in Italy.

- Training courses for gynecologists, obstetricians, and nurses.

- Workshops in local high schools to make students aware of the problem.

- Conferences for university students in the faculties of medicine, psychology, and law.

2.10 Collaboration

Collaboration with European and international universities.

2.11 Deinfibulation

The Center provides deinfibulation, which is carried out on women, with the consent of their partners, either when they get married to help avoid great pain during their first sexual relationships or during pregnancy and to give the woman a greater chance of a natural birth. Deinfibulation may also be necessary in order to carry out certain gynecological examinations.

2.12 Research

The Center provides scientific research. The results of our research into the psychosexual consequences of FGM in 137 mutilated women have been presented at the following conferences:

- 1st National Congress of the Federation of the Society of Sexology in Pisa. October 2002.
- 5th National Congress of the Italian Society of Psychology and Health in Florence, October 2002.
- 16th World Congress of Sexology in Havana, March 2003.
- 17th Conference of the European Society of Psychology and Health (EHPS) in Kos, 24th – 29th September 2003.
- 3rd Meeting of the International Society for the Study of Female Sexual Health (ISSWSH) in Amsterdam, 16th – 19th October 2003.
- 8th International Symposium on Circumcision and Human Rights: An Anthropological, Medical, Legal, and Ethical Analysis, Department of General Psychology in Padua, 2–4 September 2004.

We gave our scientific support in the "Stop FGM" meetings campaign and for many degree theses.

We made a preliminary research into defibulation: methods, motives, and consequences.

All of the above projects were held in collaboration with the Faculty of Psychology at the University of Florence.

We published some results in the Italian magazine *Sexology*, directly addressing the subject of FGM.

We studied the medical records of women with FGM to evaluate the number of caesarean sections in childbirth and, in consequence, the number of days of recovery and any ensuing complications

2.13 Preparation of Educational Material

Preparation and circulation of educational material illustrating the difference between normal female genitalia and mutilated female genitalia. This material is aimed at immigrant populations as well as students of medicine and doctors and nurses in the fields of gynecology and obstetrics.

2.14 Updating Medical Guidelines

The center keeps medical guidelines up to date in the light of new advances and research results.

2.15 Congress Organization

Organization of the second international congress, entitled *Social-Healthy Integration for Immigrants*, with the participation of the "Superior Institute of the Health."

At the first congress, a decision was made to publish a book in the Somali language addressing the dangers of FGM. This book has already been published, thanks to sponsorship and support from the Florentine local government, and it is now being translated into other languages, including Italian.

2.16 Creating a Catalogue

Establishing an open catalogue of all our publications and research to be accessed by all interested parties.

2.17 Creating a Website

The creation of a web site of the Research Center

Chapter 11

PRELIMINARY RESEARCH INTO THE PSYCHO-SEXUAL ASPECTS OF THE OPERATION OF DEFIBULATION

Saulo Sirigatti,* Lucrezia Catania,** Sara Simone,* Silvia Casale,* Abdulcadir Omar Hussen**

**Department of Psychology, University of Florence (Italy). **Research Centre for Preventing and Curing Female Genital Mutilation (FGM) and its Complications, Department of Gynecology, University of Florence (Italy)*

1. INTRODUCTION

Infibulation is the excision of part or all the external genitalia followed by stitching closed the vaginal opening (WHO, 2000). All types of female genital mutilation (FGM) have immediate and long-term complications (Cook, Dickens, and Fathalla, 2002), but long-term complications are more often associated with infibulation than with the lesser excision or *sunna*. Wheelwright (1989) states that genital mutilation results in significant reduction of sexual desire, El-Defrawi, *et al.* (2001), report that mutilated women have greater loss of desire and difficulty in reaching orgasm than intact women. Morrone (2001) describes loss of orgasm due to the amputation of the clitoris, and Rymer (2003) reports that, even if FGM has resulted in minimal physical damage, the sexual response is often decreased or absent. In the same way, Thabet and Thabet (2003) report that sexuality is markedly affected in mutilated women. Nevertheless, Ahmadu (2000) suggests that infibulation may not always have a negative impact on women's psychosexual life. Lightfoot-Klein (1989) reports that circumcised women have sexual desire, pleasure, and can reach orgasm. Megafu (1993), Okonofu, *et al.* (2002), Nwajei & Otiono (2003) say that mutilated women report that their breasts and lips are their most sexually sensitive body parts; intact women,

G.C. Denniston et al. (eds.), Bodily Integrity and the Politics of Circumcision, 123—132.

however, report that the clitoris is the most sensitive part. Catania, *et al.* (2004), in a sample of 137 mutilated women with no serious complications (who immigrated to Western countries), report that 75.3% refer to their sexual desire as intense; almost 60% report they always reach orgasm.

Nevertheless, Lightfoot-Klein (1989) states that non-orgasmic circumcised women tend not to take part in interviews about sexuality so that the results may not reflect the real situation. However, FGM is a very serious violation of the rights to physical integrity and to a complete sexual life. Thanks to suitable care for mutilated women, of which deinfibulation is a fundamental element, it is possible to improve the conditions of life for many.

Surgical deinfibulation is the procedure used to reverse infibulation. It involves the incision of the scar tissue to allow the widening of the narrowed vaginal opening. The use of surgical deinfibulation has proved effective in reducing psychosexual complications, but still tends to be little used in the developed world (Aljhadali, Amarin, Abduljabbar, 2001). Surgery is usually straightforward, but the women's expectations should be clear (The Royal Australian College of Obstetricians and Gynaecologists, 1997). In Florence, the Research Centre for Preventing and Curing FGM and its Complications has been performing this operation for more than ten years, using both traditional surgery with scalpel and CO_2-laser surgery. The deinfibulation is

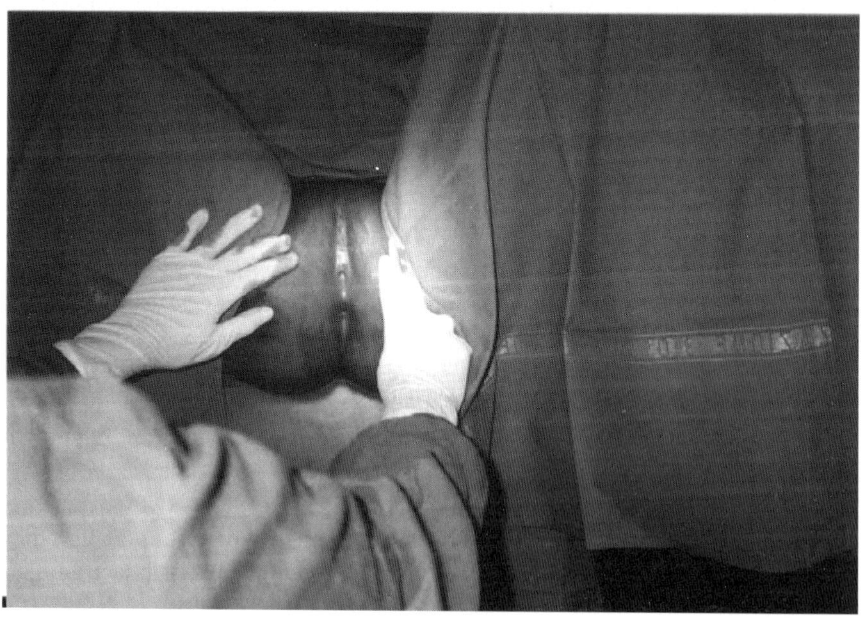

Figure 11-1. Infibulated genitals

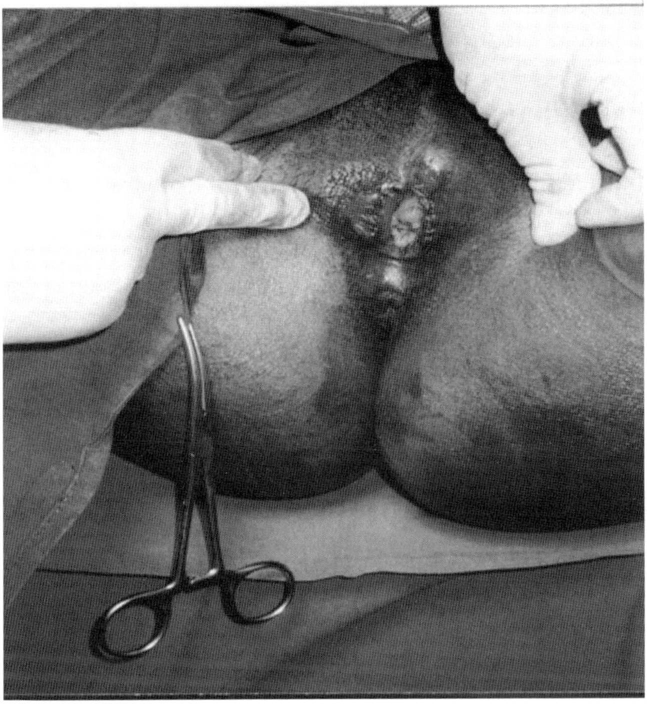

Figure 11-2. Deinfibulated genitals

performed for obstetric and gynecological reasons. It is intended to create a normal vaginal opening and to rebuild, if possible, the "normal" anatomy of external mutilated genitals. Photographs 1 and 2 show infibulated genitals before and after deinfibulation.

2. PURPOSE

To investigate socio-demographic characteristics of the samples' procedure, feelings, biomedical complications related to the infibulation, and motivations, expectations, feelings, and physical and psychological changes caused by the deinfibulation. Psychosexual answer of the deinfibulated women compared with a sample of infibulated women that have not undergone deinfibulation.

3. METHOD

3.1 Participants

•Deinfibulated Group (DG): Fifteen (15) infibulated women, immigrated from Somalia to Italy, who have undergone the deinfibulation for obstetrical or gynecological reasons at the Research Centre for Preventing and Curing FGM and its Complications in Florence. The sample has been enrolled at the center and in informal situations.

• Infibulated Group (IG): Fifteen (15) infibulated women, without serious complications, immigrated from African countries. The sample has been chosen from a sample originally wider (N =137) enrolled in the USA, in informal situations (N = 37) and in Italy, at the center, and in other places (N =100).

3.2 Instruments

A semi-structured interview was expressly created to investigate feelings about mutilation (DG and IG) and motivations, expectations, and physical and psychological changes caused by the deinfibulation (DG).

Table 11-1. Examples of Survey Questions

Topics	N°item	Item examples
Socio-demographic characteristics	31	"How long have you been living in Italy?"
Procedure, feelings, biomedical complications related to infibulation	10	"How did you feel just before the operation?"
Motivations, expectations, feelings, and physical and psychological changes caused by the deinfibulation	24	"What reasons made you decide to be deinfibulated?"
Anxiety and depression	4	"Would you define yourself as an anxious person?"
The psycho-sexual answer of deinfibulated women	26	"Do you think your sexual life has improved after dinefibulation?"
Possible alternatives to infibulation	3	"Would you infibulate your daughter?"

The group of fifteen deinfibulated women were members of the Somali community in Florence and were investigated after deinfibulation, some months after having started sexual activity. The majority (53.3%) of them worked as housekeepers, 20% were students, 13.33% worked freelance, 6.6% as employees, and 6.6% were unemployed. More than half (53.3%) of DG lived with their husband. All women were Muslim.

The psychosexual response has been investigated and compared with the psychosexual response of a group of infibulated women who have not undergone deinfibulation.

An Italian preliminary adaptation of the Female Sexual Function Index (FSFI) by Rosen, *et al.* (2000) was administered to the DG and IG groups. The FSFI is a nineteen-item questionnaire, developed as a brief, multi-dimensional self-report instrument for assessing the key dimensions of sexual function in women in the most recent four weeks. The FSFI investigates six domains: desire, arousal, lubrication, orgasm, satisfaction, and pain. Women freely agreed to answer the questionnaire. They were contacted by a specialized, trained gynecologist renowned, in the Italian and European community of immigrants, for his work in this field.

4. RESULTS

Women's average age of the DG was 28.8 ± 4.9 years; and, for the IG, it was 29.2 ± 3.7 years. Average age of DG at the moment of the infibulation was 9.42 ± 26.88 months. "Where was the operation performed?" Among the DG, 33.33% were in hospital, 33.33% in the house of the operator, 20% in their own homes. 6.6% in a neighbor's house, and 6.6% do not remember. About the person who performed the operation: 40% of the DG refer to a doctor, 40% to a nurse, 13.33% to a local woman who usually performs the mutilation, and 6.66% do not remember. According to 66.66%, anesthetics and antibiotics were used to relieve pain and prevent infections.

How did you feel in the days after the operation?

In accordance with El-Gibaly (2002), infibulated and deinfibulated women expressed ambivalent feelings and conflicting childhood memories about their mutilation.

P

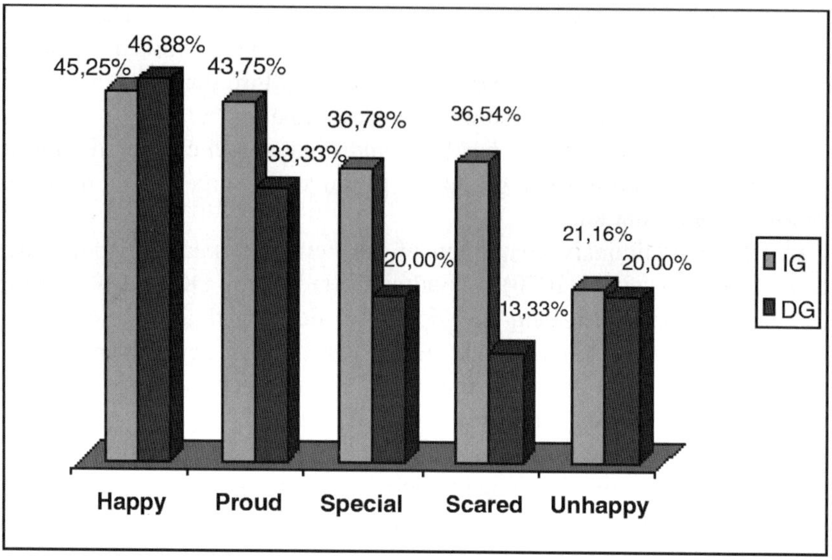

Figure 11-3. How did you feel in the days after the operation?

Table 11 -2. Did you have any complications in the years after the operation?

Description	n*	%	n*	%
None	2	13.33	0	0.00
Urinary infection	2	13.33	n.a.	n.a.
Urinary retention	1	6.66	2	13.33
Menstrual pain	12	80.00	9	60.00
Pelvic infection (pain, temperature)	0	000	n.a.	n.a.
Pain during sexual intercourse	7	46.66	4	26.66
Huge scars or cysts in the scar	2	13.33	0	000
Cysts and abscesses	2	1.33	n.a.	n.a.
Difficulty in getting pregnant	2	13.33	0	000

The deinfibulation was performed because of medical reasons (46.66%), for having the first sexual intercourse (40%), for improving sexual life (46.66%), for reducing the probability of a difficult childbirth (13.33%).

Table 11-3. After the deinfibulation, what changes in your body did you most appreciate?

Description	n*	%
Urinating more easily	7	46.66
Having a more abundant flux	4	2.66
Having less menstrual pain	7	46.66
Having less painful intercourse	11	7.33
Having no more infections	2	13.33

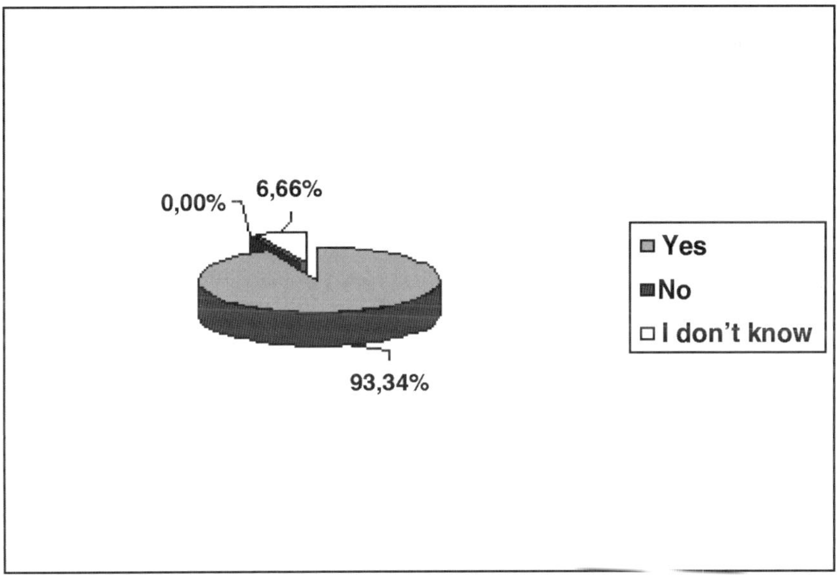

Figure 11-4. If you could go back in your life, would you choose to be deinfibulated?• Psychosexual answer of the deinfibulated women (53.33% of the DG were sexually active).

Do you think that your sexual life improved after deinfibulation?

Can you reach orgasm with penetration?

Table 11-4. Psychosexual Aspects

Item	SS effect	Df effect	MS effect	SS error	Df error	F	p
Pain during penetration	4.03	1	4.03	38.27	28	2.95	0.10
No desire	1.63	1	1.63	23.33	28	196	0.17
Sex repugnance	0.03	1	0.03	7.47	28	0.13	0.73
Sex without desire	0.30	1	0.30	25.07	28	0.34	0.57
Sex with desire	0.03	1	0.03	29.33	28	0.03	0.86
Simulated orgasm	0.03	1	0.03	9.33	28	0.10	0.75
Sex importance	1.20	1	1.20	11.60	28	2.90	0.10

Table 11-5. FSFI Domains: DG and IG

Domain	**S effect**	**Df effect**	**Ms effect**	**SS error**	**Df error**	**F**	**p**
Desire	3.33	1	3.33	139.87	28	0.67	0.42
Arousal	140.83	1	140.83	1005.87	28	3.92	0.06
Lubrication	34.13	1	34.13	1241.87	28	0.77	0.39
Orgasm	149.63*	1*	149.63*	696.67*	28*	6.01*	0.021*
Satisfaction	13.33	1	13.33	431.47	28	0.87	0.36
Pain	563	1	5.63	1033.33	28	0.15	0.70

5. CONCLUSION

In accordance with Morrone (2001), women have been infibulated in hospital by doctors, and medicine (antibiotics and anesthetics) generally has been used. Women have different feelings about their condition of being infibulated (happy, proud, scared, and unhappy). In accordance with Obermeyer (1999), no serious short- or long-term complications have been observed. In accordance with Amhadu (2000), this study suggests that deinfibulation has no definite negative impact on women's psychosexual life. Women were all satisfied with deinfibulation; the complications of infibulation have been improved by deinfibulation. In accordance with Lightfoot-Klein (1989) and Catania, *et al,* (2004), nearly all deinfibulated women can reach orgasm with penetration.

From the Analysis of Variance between the sample of deinfibulated women and the sample of infibulated ones, there is a significant difference in the orgasm scale of FSFI. On the contrary, no significant differences were found in the items of the interview about psychosexual aspects.

All women of DG and IG were sexually active. Before the deinfibulation, 53.33% of DG had already had sexual intercourse by penetration. The quality of the sexual life of DG after the deinfibulation improved in 46.66%, and 93.33% of DG could have orgasm, only 6.67% could not.

All women agreed to answer the questions about orgasm: their answers were unambivalent because they described the psycho-physical effects that connoted the greatest moment of pleasure/arousal that they have defined as orgasm (involuntary pleasurable rhythmic contractions of the vagina, pulsations of the internal genitals, feeling of warmth all over the face and body, crying out of uncontrollable words or sounds, a complete abandoning of the body, feeling of loss of control, feeling of exploding or melting). Our research will complete the results with laboratory data (e.g., vaginal blood flow, nuclear magnetic resonance of genitalia, electromyogram, ultrasound scan of pelvic muscles, and color Doppler) in the group of infibulated and defibulated women we investigated.

REFERENCES

Abdulrahim AR, Aljhadali, EA, Amarin, ZO e Abduljabbar, ZO (2001). The use of intrapartum defibulation in women with female genital mutilation. British Journal of Obstetrics and Gynaecology, 108, 949-951.

Amhadu F (2000). Rites and Wrongs: An Insider/Outsider Reflects on Power and Excision. In B. Shell-Duncan, and Y Hernlund (Eds.), Female Circumcision in Africa: Culture, Controversy, and Change, 52, 283-312. London: Lynne Rienner Publishers.

Catania L, Baldaro VJ, Sirigatti S, Casale S (2004). Indagine preliminare sulla sessualità di un gruppo di donne con mutilazioni dei genitali femminili (MGF) in assenza di complicanze a distanza. Rivista di Sessuologia, vol. 28, no. 1. Gennaio-Marzo 2004. pp. 26-34.

Cook RJ, Dickens BM, and Fathalla MF (2002). Female genital cutting (mutilation/ circumcision): ethical and legal dimensions. International Journal of Gynaecology and Obstetrics, 79, 281-287.

El-Gibaly O, Ibrahim B, Mensch BS, Clark WH (2002). The decline of female circumcision in Egypt: evidence and interpretation. Soc Sci Med, 54 (2), 205-220.

El-Defrawi MH, Lotfy G, Dandash KF, Refaat AH, and Eyada M (2001). Female genital mutilation and its psychosexual impact. Journal of Sex & Marital Therapy, 27(5), 465-473.

Lightfoot-Klein H (1989). The Sexual Experience and Marital Adjustment of Genitally Circumcised and Infibulated Females in the Sudan. The Journal of Sex Research, 26(3), 375-392.

Megafu U (1993). Female Ritual Circumcision in Africa: An Investigation of the Presumed Benefits among Ibos of Nigeria. East African Med. Journal, 40 (11), 793-800.

Morrone A (2001). Mutilazioni genitali femminili: Un problema nuovo e antico. Esperienze Dermatologiche. 3 (4):413-430.

Nwajei SD and Dtiono AI (2003). Female genital mutilation: implications for female sexuality. Women's Studies International Forum, 26 (6), 575-580.

Okonofu FE, Larsen U, Oronsaye F, Snow RC, and Slanger TE (2002). The association between female genital cutting and correlates of sexual and gynaecological morbidity in Edo State, Nigeria. British Journal of Obstetrics and Gynaecology, 109(10), 1089-1096.

Rymer J (2003). Female Genital Mutilation. Current Obstetrics and Gynaecology, 13(3), 185-190.

Thabet SM and Thabet AS (2003). Defective sexuality and female circumcision: the cause and the possible management. The Journal of Obstetrics and Gynaecology Research.

Wheelwright J (1989). Female circumcision comes under the knife. New Africa, 265, 26.

WHO. Management of pregnancy, childbirth and the post-partum period in the presence of female genital mutilation: report of a WHO technical consultation, in: WHO. Female Genital Mutilation: a handbook for frontline workers. WHO, Geneva, 2000.

Chapter 12

ADDRESSING FEMALE GENITAL MUTILATION IN GERMANY

The Work of the Women's Rights Organization Terre des Femmes and the Situation in Germany

Petra Schnuell, Gritt Richter, Claudia Piccolantonio
Terre des Femmes

Abstract: Terre des Femmes is a women's rights organization, whose name translates as "Women's Earth." Terre des Femmes is committed to the problem of female genital mutilation in Germany.

1. INTRODUCTION

The German public was first confronted with the problem of female genital mutilation (FGM) in 1977, when articles on the subject were published in the news magazine *Der Spiegel*, as well as in the excellent publication of the feminist movement, *Emma*. In 1978, a very detailed and blunt article by Fran Hosken was published in *Courage* magazine. People who read the magazine were shocked, uncomprehending. Nobody could believe it, moreover, nobody wanted to believe it. People would have preferred to be spared hearing about the issue. This is why the feminists who wanted to campaign for the abolition of female genital mutilation were not welcomed with open arms. At first, their voices fell, for the most part, on deaf ears.

2. WALKING A TIGHT ROPE

One major reason for the lack of interest was the blunt and rather simplistic way in which the issue was presented. This could be why many

G.C. Denniston et al. (eds.), Bodily Integrity and the Politics of Circumcision, 133—140.

people baulked at the subject: it was less a matter of ignorance and more a matter of self protection. The feminists wanted to stir up people's awareness as powerfully as possible in sincere solidarity with the girls and women affected. For the mass media, however, the issue was tailor made — and apparently still is — for boosting circulation and viewing figures with gore and sensationalism. This counterproductive and sensationalist reporting made the work of FGM activists even harder. Now they also had to fight deliberate disinformation, burgeoning resentment, racist tendencies, and voyeurism. Early public relations work on the issue was like walking a tightrope. Gradually, with time, this was taken in stride. A dynamic learning process was set in motion, and this is reflected in particular in the work of Terre des Femmes. Terre des Femmes is concerned with presenting the issue in all its complexity but without playing it down. It is not a neutral issue; female genital mutilation is viewed as a severe violation of human rights. In the case of doubt, however, priority goes to the dignity, interests, and opinions of those directly affected by FGM. Problems that are "foreign" in one society — one could even say "alienating" — can only be discussed and questioned in a dialogue of equals.

3. A DIFFICULT DIALOGUE BEGINS

The dialogue between the European and black African activists, who joined the anti-FGM movement at a later date, was very difficult and, at first, recriminatory. African activists felt they were not properly represented and that their voices were not being heard. In 1999, during this difficult period, a voyeuristic poster campaign was launched against FGM, which led to a public stigmatization of African women living in Germany and helped increase the already latent racism. This campaign widened the gap between European and black African activists. Another disastrous result was that, now, the "white" organizations were apparently no longer qualified to deal with the issue. This had a serious impact on Terre des Femmes. The wounds and resentment were difficult to heal, and we only really have overcome them in the past two years.

4. A KEY ISSUE

FGM has always been a key issue at Terre des Femmes, ever since its creation in 1981. Since then, we have become the foremost FGM information center in Germany. The type and scope of the incoming queries are as varied as the origins of the senders themselves. They range from individuals, who are interested in the issue for personal or professional reasons, to representatives of the media or of various organizations and governmental offices, and to those affected by FGM themselves. In 1995, women volunteers from all over Germany got together to form an FGM study group under the auspices of Terre des Femmes. This group meets twice a year to share information, to network, and to plan joint activities. In 1997, the Terre des Femmes office recruited a full-time expert. With her help, work that had previously been performed by volunteers became more professional.

5. PROVIDING INFORMATION

Our work to abolish female genital mutilation is based first and foremost on information. We draw people's attention to the problem, lobby politicians, educate, and try to drum up the solidarity and support of as many people as possible to make our efforts even more effective and intensive. To do so, we use all the means at our disposal. For example, we regularly publish articles on the issue in the Terre des Femmes magazine, *Menschenrechte für die Frau* (Human Rights for Women), with reports on new developments or grassroots projects sponsored by Terre des Femmes. In addition to flyers for public relations work, we have also published a brochure for women immigrants entitled "We want to protect our daughters." We based this on a French brochure that has been used successfully in prevention work for many years. This collaboration gave us valuable experience and, at the same time, demonstrated how two European countries work together to abolish female genital mutilation. The brochure is aimed at prevention in Germany, informing women, German legislators, and international conventions. The brochure is available in six languages:—German, English, French, Arabic, Somali, and Kiswahili.

We also have published two books on the subject, with contributions from African, European, and American experts on FGM to allow people who are interested but have no previous knowledge of the subject to get the facts. The first book was published in 1999 and was immediately sold out. The second book, published in 2003, is also in demand. Less and less literature is available on the subject in German. This is all the more surprising since relatively large numbers of people are now interested in it. We can thank Waris Dirie's book, *Desert Flower,* for a major part of the interest in the subject. It was featured on the best-seller list for a long period of time and had a very broad impact.

Our website is updated regularly. There, besides plenty of information and numerous links, are listed the addresses of centers for those affected by

FGM and, unique in Germany, international guidelines for health workers. We have organized several touring exhibitions, most of them showing photographs, but one showing paintings, entitled "Female Genital Mutilation: Nigerian Artists Speak Out." Another notable example is our short film, "A Very Special Day," which has no dialogue. It tells the story of an African mother, living in Germany, who has made all the arrangements for the circumcision of her daughter. Suddenly, she remembers the pain of her own circumcision and, in the end, spares her daughter the operation. Terre des Femmes placed great importance on the fact that the film, which was produced by the Munich Film Academy, should be neither shocking nor voyeuristic. The filmmakers met these requirements. To date, the film has been featured on the supporting bill in over 150 cinemas in more than 100 towns and cities. By Germany's standards, that is a significant social sponsoring.

We also hold numerous talks on the issue, organize seminars, workshops, vocational continuous education courses, and panel discussions. We work with schools and universities, youth welfare departments, midwifery schools, and with the Federal Office for the Recognition of Foreign Refugees. We frequently collaborate with other organizations and especially welcome the chance to work with activists from Africa. We continuously train people in expert meetings and are part of national and international networks.

6. LOCAL INITIATIVES

We support local initiatives, sponsoring three grassroots projects in Burkino Faso, Tanzania, and Kenya. All three projects were initiated by local people and are managed by them. Direct involvement is not part of the project philosophy of Terre des Femmes because nobody is better qualified to enlighten people than those concerned with their own ethnic group. This is why we restrict our work basically to fund-raising or passing on information, if it is requested.

Luckily, women's rights organizations such as Terre des Femmes today are no longer alone in their commitment. Now, all human rights activists in Germany, male and female alike, agree that female genital mutilation should not be portrayed as a cultural tradition worthy of conservation but that as a violation of human rights specific to women that must be abolished.

The international movement against FGM for decades was impeded by accusations of cultural imperialism and "Eurocentricity." The first opponents of the practice in Germany, the feminists, were thereby silenced. In the late 1970s, so-called "progressive" circles attacked these early activists as racist and "Eurocentric." The subject continued to be sensitive in Germany.

When Terre de Femmes took up the issue in 1981, it too was attacked. This situation changed little up to the mid-1990s. Public relations work proved difficult and bore little fruit. Even the outcome of the World Conference on Human Rights in Vienna in 1993, which, for the first time, firmly anchored FGM in the human rights debate, bore no results for many years. Despite Terre de Femme's 1997 estimate that at least 21,000 women in Germany were affected, the message was always the same: Africa is a long way away. There are no publicly known cases. Therefore, there is no need to act.

Only in 1997 did the issue gain greater public notoriety, with the hearing initiated by the Bündnis90/Green party in collaboration with Terre des Femmes, entitled "A pain that touches the soul." One year later, the parliamentary factions for the first time expressed an opinion on the subject, judging that female genital mutilation constituted a serious violation of human rights and severe bodily harm.

From then on, society in general began to move and the issue became more widely known. But the real breakthrough among the general public came with Waris Dirie's autobiography, which was published in Germany in 1998 under the title *Desert Flower* and was read by people of all age groups. Moreover, in March 1999, a report on German television shook many people and put the whole issue in a new light. It was reported that an Egyptian doctor, based in Berlin, was prepared to practice FGM on young girls for the price of 610 euros. The investigation was abandoned due to lack of evidence and also because police intervention came too late. Public interest, however, was aroused.

7. FGM A NEW ISSUE

Female genital mutilation only became an issue in Germany in 1997, which explains why Germany is lagging behind other Western nations on FGM. FGM gradually gained the attention of German society. This resulted in the creation of other associations, which, like Terre des Femme, took up the issue. The most notable of these include FORWARD Germany and INTACT. With Terre des Femmes, they form the core of the anti-FGM movement in Germany. Other associations that are working to combat the practice include DAFI, G.R.A.F., and "stop mutilation."

Each of these associations has chosen to focus on different methods in their struggle against FGM, which has created an important synergy effect. Terre des Femmes concentrates primarily on public relations work in Germany, while INTACT places more importance on supporting projects in

hospitals in order to make the procedure safer and to respect the traditions of those populations who for centuries have undergone the intervention.[8]

The National Bioethic Committee (NBC), which recently addressed the issue of ritual circumcision,[9] emphasized the ethical problems with female genital mutilation and briefly discussed male circumcision. According to the NBC, the major concern was whether the circumcision of a minor was legal. The NBC also debated whether ritual male circumcision should be performed under the National Health Services (NHS). The 1998 resolution addressed prophylactic male circumcision and male circumcision carried out for other reasons, which were deemed unjustifiable. The resolution allows parents or guardians to choose circumcision on behalf of children, in accordance to the articles eight, nineteen, and thirty of the Italian Constitution, which guarantee freedom of religion. The resolution referred to articles 2.1 and 26.1 of the *Agreement Between the Italian State and the Hebrew Communities*, which acknowledged the right to perform, in any collective or individual way, the rites of Judaism. It was resolved (albeit incorrectly) that male circumcision — apart from the fact that it is irreversible — if correctly performed, is alleged not to alter male sexual and reproductive function. Consequently, the NBC seems to have adopted the position that male circumcision falls outside the scope of 'acts dangerous to the individual' and is therefore legal. Nevertheless, ritual male circumcision cannot be performed under the NHS. All the parties to the agreement also concurred that male circumcisions must be performed by a medical doctor. Some NBC members, however, claimed that ritual circumcision of newborn males (but not of infants or adults) could be performed by competent nonmedical religious personnel. In any case, those who perform male circumcision are responsible for its correct execution, asepsis, hygiene, and postsurgical care.

The lack of unambiguous legislation regarding ritual male circumcision enabled fraudulent use of NHS funding. The press has reported instances where circumcisions that were really performed for ritual reasons were falsely labeled as being performed for "therapeutic" reasons in order to have them performed gratis under the NHS.[10]

Because demographic and statistic information about the prevalence and incidence of male circumcision in Italy has never before been established, the following study was performed, and its findings are summarized below.

2. MATERIALS AND METHODS

Using discharge documents collected from Italian medical clinics for the years 1999, 2000, and 2001, we were able to make use of diagnosis-related groups (DRG) for our study. DRG 342 (male circumcision performed on subjects older than seventeen years) and DRG 343 (subjects younger than eighteen years) were considered.[11] We identified and noted the type of medical facility that generated the discharge documents (state, private, research institute, etc.). The type of hospitalization (out patient or overnight) was ascertained. Secondly, for therapeutic male circumcision, codes for IDC-9-male circumcision were analyzed (exuberant prepuce, phimosis, and balanoposthitis).

We also conducted a survey of American and Italian workers of the United States Air Force base, *Caserma Ederle*, located near Vicenza. This was done to ascertain whether Italians living near the American military base might have been influenced by the presence of circumcised American soldiers, as happened in South Korea. In Korea, however, mass circumcision was imposed on the Korean population by American military rulers.[12] The American community of Vicenza is composed of about 2500 soldiers, 600 civilians, 4000 relatives. Approximately 700 Italian civilians are employed at the military base. To investigate the phenomenon, two semi-structured interviews were conducted with the American pediatrician of the military base and with an Italian pediatric surgeon at San Bortolo Hospital, also located in Vicenza.

3. RESULTS

The main results are shown in graphs 1 and 2. In general, male circumcision is carried out in health facilities and, at least until the age of fifteen, in regular hospitals. From 1999 to 2001, the number of therapeutic male circumcisions increased from six percent in the year 2000 to nine percent in the year 2001, with respect to year 1999. The number of interventions carried out in day hospitals increased, compared to regular hospitals, which decreased). The increment passed from more than twenty-eight percent (2000) to fifty-one percent (2001), compared to 1999.

Graph 3 depicts two age peaks in interventions: 5-14 and 25-44.

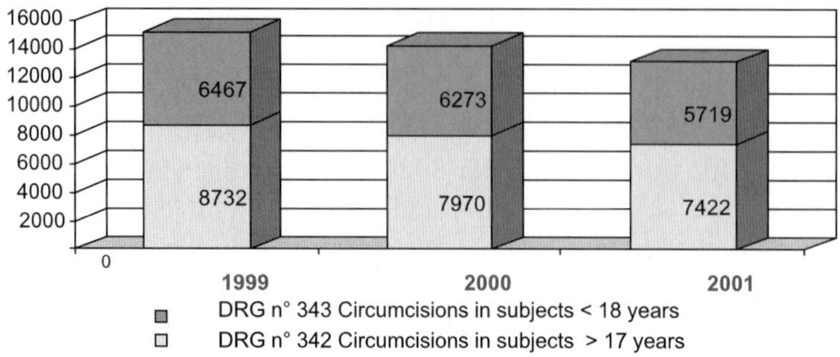

Figure 13-1. Graph 1 - Circumcisions carried in regular hospitals from 1999 to 2001.

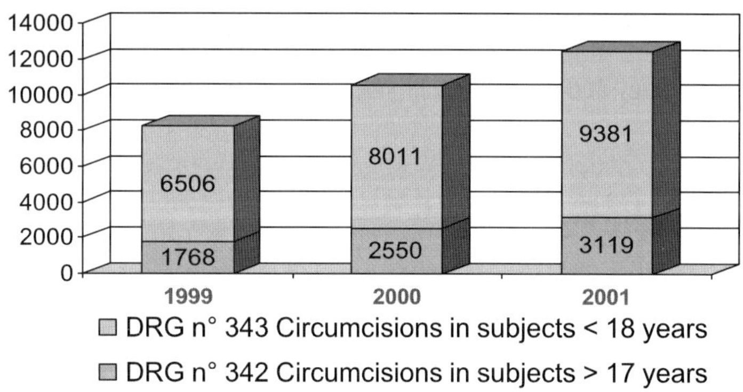

Figure 13-2. Graph 2 - Circumcisions carried out in day hospital from 1999 to 2001.

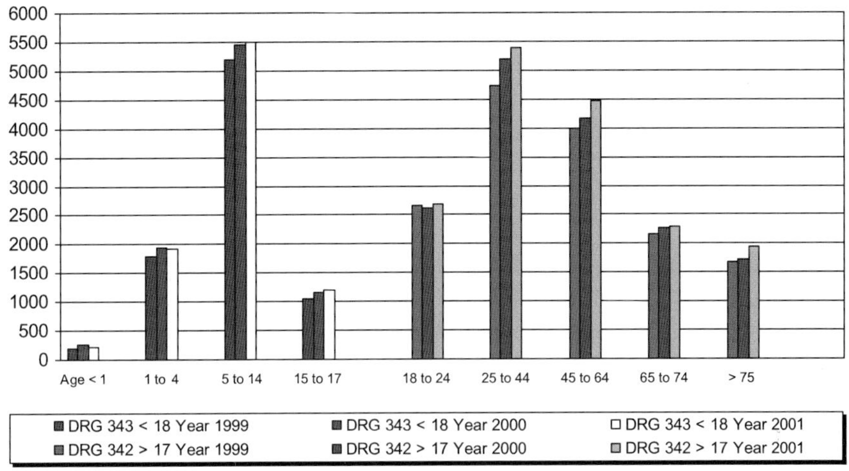

Figure 13-3. Graph 3 - Age classes in which circumcision was carried out in hospitals – Data from 1999 to 2001.

4. DISCUSSION AND CONCLUSIONS

The increase of male circumcision from 1999 to 2001 may be a result of recent waves of immigration from Islamic countries. Immigrants to Italy (year 2000) primarily come from Morocco (n = 127,993), Albania (n = 99, 230), Tunisia (n = 43, 359), Senegal (n = 35, 270), Egypt (n = 27, 142), Bangladesh (n = 13, 803), Algeria (n = 10, 671), and Somalia (n = 4, 482).[13]

Concerning the two age groups in which male circumcision is generally performed, the first peak could be a result of pediatric check-ups carried out at school that allow identification of phimosis previously undetected. The second peak may reflect a diagnosis of secondary phimosis in adults. It is possible that within the first age group some ritual male circumcisions are fraudulently carried out under the guise of treating "congenital phimosis," as the press recently reported in the case of a surgeon of Maghreb origin, practicing in Northern Italy, who performed ritual circumcision on Muslim children. The sum of 1,033 euros was charged to the National Health Service for each circumcision under the ruse that the operations were performed for the treatment of "phimosis." The same article reported that, in 1998, in the mosque in Segrate, an Egyptian doctor performed ritual circumcision on males and, supposedly, infibulation on girls.[10]

Apart from the number of male circumcisions carried out in medical clinics, it is difficult to estimate the number of circumcised males in Italy due to the increasing number of illegal immigrants who often cross the Italian borders in order to pass into northern European countries. It is also difficult to estimate the number of ritual circumcisions carried out on non-Hebrews because illegal circumcisions are often performed clandestinely in mosques at a cost of 120 to 150 euros.[10]

The Vicenza survey revealed that Italian and American doctors that we interviewed displayed two contrasting positions. The American doctor was a strong supporter of routine male circumcision. In his opinion, the alleged benefits of the surgery outweigh the harm. The Italian surgeon held the opposite conviction, justifying his assertions on ethical and cultural grounds. Both of them claimed that Italian residents are uninfluenced by this American cultural phenomenon. Informal contacts with Italians working with Americans at the military base revealed that most were unaware that routine male circumcision was widespread among the Americans.

The attitudes of Italian urologists about male circumcision differ. An analysis of views expressed in the popular press revealed that many urologists believe that, after age four, male circumcision is recommended in cases of phimosis to prevent complications. It is also believed that male circumcision should be offered to the immigrant community in order to reduce the risks of infection and hemorrhage. The Ministry of Public Health, however, does not include male circumcision among the "essential levels of assistance" interventions, even though some public health clinics in 2004, in areas with a large immigrant Muslim population, permitted ritual male circumcision at low cost, resulting in a political crisis.[14]

Due to cuts in the healthcare budget, the Ministry of Health declared that many interventions previously performed in regular hospitals would now be carried out in day hospitals. This could explain the substantial increase of male circumcisions carried out in day hospitals from 1999 to 2001.

As in all other European countries, male circumcision is uncommon in Italy, a country with a strong Catholic heritage that values and promotes the integrity of the human body. The Hebrew minority in Italy is very small, as was the Muslim minority until a decade ago. Muslims, however, are now increasing in number, and this requires that the phenomenon be closely monitored in order to protect children from unwanted ritual surgeries.

Recently, the Justice and the Social Affairs Commissions of the Italian Parliament held debates over female genital mutilations (FGM), resulting in a legal proposal.[15] The eleventh Permanent Commission voted favorably to ban FGM, but required removal of the word "female." The aim was to protect both sexes equally against genital and sexual harms and mutilations, as the Italian Constitution affirms equality between the sexes.[16] While the popular media has done much to publicize the horrors of female genital mutilation, it has not yet done so for male circumcision.

The present study is on-going and will next survey and document the attitudes of Italian urologists toward male circumcision and present a thorough summary of the relevant statements made by the various bioethics committees working in Italy.

REFERENCES

[1] Celsus A.C. (1971) De Medicina. William Heinemann Ltd. Ed., London.

[2] Falloppius G. (1600) Tractatus de Decoratione – Tome II, Joann Petri Maphaei Ed, Frankfurt.

[3] Fabritius ab Acquapendente I. (1619) De chirurgicis operationibus. In: Operationes Chirurgicas in duas partes divisas. Apud Paulum Meglieim Ed., Venezia.

[4] Mantegazza C. (1885) La questione antisemita. Fanfulla della domenica, n. 38-39.

[5]. Lombroso C. (1894) L'antisemitismo e le scienze moderne. Torino.

[6] Savaresi F. (1802) Memoires et opuscules physiques et médicaux sur l'Egypte. Paris X ed. Paris.

[7] Tomès R. (1895) Della Circoncisione. Tipografia dell'Annunzio. Firenze.

[8] Andretta E. (2000) Circoncisione, rispetto di una cultura. Il Gazzettino, Inserto Salute, 13 Marzo 2000, page 14.

[9] Comitato Nazionale di Bioetica (25 Settembre 1998). La Circoncisione: Profili Bioetici. Governo Italiano. Presidenza del Consiglio dei Ministri, Roma.

[10] Various Authors (2000). Circoncisioni a carico del servizio sanitario. La Padania, 28 luglio 2000.

[11] Ministero della Salute. Dipartimento della Programmazione (1997). Classificazione della malattie, dei traumatismi, degli interventi chirurgici e delle procedure diagnostico-terapeutiche. Istituto Poligrafico e Zecca dello Stato. Libreria dello Stato, Roma.

[12] Dunsmuir WD, Gordon EM. (1999). The history of circumcision. BJU International Ed: 83(1): 1-12.

[13] www.caritasroma.it

[14] Il Gazzettino. 30 luglio 2004.

[15] Report in shorthand of: Atti Parlamentari – Camera dei Deputati. XIV Legislatura – Discussioni. Seduta del 29 marzo 2004.

[16] Atti Parlamentari - Camera dei Deputati – XIV Legislatura – Disegni di Legge e Relazioni – Documenti. Proposte di Legge n. 150. N. 150-3282-3867-3884-4204-A. Parere della XI Commissione Permanente, 8 aprile 2003.

Chapter 14

GENITAL INTEGRITY AND GENDER EQUITY

J. Steven Svoboda
Attorneys for the Rights of the Child

Abstract: The struggles to protect male and female genital integrity may be analyzed as parts of larger struggles for men's and women's rights. While genital integrity still lacks binding legal safeguards, individual victories have been won. A number of important gender equity cases have been decided by the United States Supreme Court affirming men's right to equal treatment. Several illuminating parallels link genital integrity and gender equity. Like genital integrity, gender equity is fundamentally a civil rights movement. Both topics seem strange at first but ultimately derive from basic principles. Both causes are relatively unfamiliar to the broader legal community, not to mention the general public. Both can be rendered easily understandable through parallels with familiar topics. Genital integrity may *temporarily* be less legally robust. Both movements may be on the verge of some major breakthroughs. The interconnections between gender equity and genital integrity are growing in importance. Awareness of the gender equity movement should inform our activism regardless of our personal views. Both movements are relatively neglected at this particular historical moment. The very resistance such discussions raise may represent the barriers faced by activism in gaining support for its male-protective cause.

1. INTRODUCTION

American society evidently finds it very difficult to accept the profound importance of protecting the genital integrity of, at minimum, all individuals below the age of consent. Most blatantly, male circumcision is performed on well over a million US babies each year,[1] while female circumcision has been a federal crime for nearly a decade.[2] A number of different approaches have been used by commentators in trying to get to the root of the difficulty. One issue, bearing many potentially instructive lessons, has been largely

G.C. Denniston et al. (eds.), Bodily Integrity and the Politics of Circumcision, 149—164.

neglected in activist literature. We may learn much from comparing the movement for genital integrity with the movement for genuine gender equality. The two forms of activism bear a complex relationship to each other, as multifaceted as the many fascinating parallels connecting genital integrity activists and gender equity activists.

In referring to activists for genuine gender equality, I am regrettably not primarily referencing mainstream feminism in its current manifestation. Feminism was undeniably dedicated for many decades to securing gender equality for both men and women (which is a somewhat redundant phrase in any event). The feminist focus in the seventies and early eighties on passing an Equal Rights Amendment, coupled with a failure in more recent times to continue pursuing this reasonable legislation, merits inquiry. To some commentators, mainstream feminism now appears more focused on achieving privileges for women (or, in some cases, for "women and children"). At the same time, today there remain a large number of women who identify themselves as feminists and support men's equality as well. Entire branches of feminism such as "equity feminism" concentrate on advancing issues of genuine gender equality. Leading proponents include Christina Hoff-Sommers and the organizations International Women's Forum (IWF) and Women's Freedom Network (WFN).

2. GENDER EQUITY AND GENITAL INTEGRITY ACTIVISM

Personally, I find that I am spending a good percentage of my life fighting struggles that, in a better world, would never have been necessary, and also struggles whose lack of widespread endorsement never fails to surprise me. What could more clearly coincide with common sense than the principle that males as well as females should, in the absence of extremely compelling and exceptional reasons militating otherwise, retain the genitals that are their birthright? Perhaps only the even more basic idea that males, as well as females, should be entitled as far as possible to freedom from discrimination based on the nature of their genital equipment.

Gender equity envisions and struggles for a world free of discrimination against women in employment and free of discrimination against men in custody and child support cases. Gender equity activists fight for a future in which female health and male health are equally cherished and equally protected, so that funding flows amply to stop both prostate cancer and breast cancer, males and females are equally represented in clinical trials, and society addresses itself immediately to reversing men's six fewer years of life expectancy. (Already, New Hampshire has opened the world's first men's affairs commission run under government auspices, though for the present it does not receive a dime of government money.) Those who strive to promote gender equity imagine a country in which criminal sentencing does not assign harsher sentences to men and in which males are not subject to a discriminatory military draft.

Genital integrity activists fight to protect men's (and of course women's) rights against amputation of a functional body part without their informed consent. The struggles to protect genital integrity may be analyzed as parts of larger struggles for men's and women's rights.

Yet, both the genital integrity and gender equity campaigns, while seemingly reflecting the most obviously meritorious of principles, are nevertheless paradoxically ostracized from widespread acceptance by a variety of factors. These barriers to acceptance include the reigning paradigm under which men's and women's discrimination are viewed differentially, lack of exposure due to historically meager press coverage (though as we shall see, currently media attention to both genital integrity activism and gender equity is growing), and court reluctance to affirm issues not yet socially approved. Also, both movements quite simply may cause some level of discomfort to individuals, institutions, and, indeed, society itself. Genital integrity partakes of sex, religion, psychological denial, medical procedures, parental denial, and a variety of other uncomfortable issues. Men's rights (if we use this somewhat misleading terminology) is almost as unappealing, striving to affirm the need for equal treatment of a sex that is seen as historically dominant and currently privileged in many or most aspects of human life.

A "hold back the floodgates" mentality may also be at play. If males are "in" as rightful claimants, the thinking may go, then nobody is "out" any more. The impulse to define one's efforts in terms of insiders and outsiders is natural; it may be harder to visualize an enemy when everyone is potentially a victim.

The same forces that separate the genital integrity and gender equity movements from broader success also separate the two movements from supporting each other. For many if not most of the reasons discussed above, activists for genital integrity often view other claims by men with a certain suspicion. Probably playing into this perspective are fears of message dilution, simple logistical inability to take on more than one struggle, submission to the reigning paradigm in which men's difficulties are not as easily acknowledged, and so on.

Similarly, many gender equity activists are uninterested in getting involved in such a seemingly outlandish and strange issue as male circumcision. Here, as well, we may find fear of message dilution, logistical limitations, and a submission to an equally pervasive dominant perspective in which circumcision is not harmful (or, at least, not very harmful), has medical benefits, and in any event is a religious sacrament so must not be challenged. One other factor in the rejection of genital integrity as a proper concern is conscious or unconscious homophobia, whose powerful role among heterosexual American males should not be underestimated. Male genital integrity activists may be perceived as likely to be gay due to their evident interest in male genitalia.

One further parallel is that discrimination against males and violation of male genital integrity are both rendered more acceptable by focusing on supposed gender differences that justify differential treatment that hurts males while sparing females. The validity of these distinctions tends to quickly evaporate upon careful inspection.

Finally, we are all in a common struggle to advance ourselves as individuals and as a culture. Anti-male discrimination, as with forced military service for males, can often transmute into anti-female discrimination, as with prohibitions against female participation in certain sectors of the military. Similarly, the same arguments used to buttress alteration of male genitals in one part of the world support female genital cutting in another part. As recently as the 1970's, female genital mutilation was being performed and recommended in reputable medical journals by physicians within the United States![3] Males have wives, daughters, mothers, and friends who dearly care about them, so we deceive ourselves if we imagine that harm to one gender does not ultimately impinge on the other. We are all connected.

2.1 Definitions, Nomenclature, and Purposes

It has been said that the result of a debate often hangs on how it is framed. There are few movements in this world for which this is so true as genital integrity and gender equity.

Problems with definitions and nomenclature exist in both movements that reflect society's confusion over the issues.

Are we against circumcision? Or, are we for genital integrity? Do we oppose so-called non-religious circumcision, all circumcision, or what? Are we against adult circumcision?

A similar set of questions arises for gender equity. Are we for men's rights, whatever that might mean? Are men essentially to blame for at least some of their problems and, if so, do we need to do anything about them? What if the pot is limited and supporting greater rights for men in practice involves fewer rights for women?

There are as many answers to these questions as there are activists. Part of the reason for the obscurity and struggles of both movements is that society lacks the cognitive tools to develop its own commonplace understanding of the answers to these questions. Other social causes are more generally understood. Whatever one's feelings about them, people grasp the positions, for example, of so-called pro-life and so-called pro-choice activists.

2.2 A Personal Note

I became interested in genital integrity first, back in 1989, while I was still a law student, and also have been interested in all forms of gender equality. The same factors of seeking fairness and justice motivate me in both areas. In my life, the two have gone almost hand in hand. I personally am first and foremost a lifetime activist for fairness and equality in all aspects and all phases of life. I also considered myself a feminist for many years and was dismayed when I found myself forced to abandon that affiliation. When I first attended law school so I could gain more tools to fight for social justice, I never dreamed that someday I might be working on behalf of males, let alone in two different pursuits!

My dedication has rarely wavered but has largely increased since I first became involved in each issue. Currently, I serve as a Board Member and the Public Relations Director of the National Coalition of Free Men (NCFM), the oldest and largest membership organization in the world devoted to promoting genuine gender equity.

2.3 A Note on Some Leading Gender Equity Organizations

Interestingly, the National Organization of Women actually conferred an award on NCFM in 1980 for its report entitled "Ties that Bind: Price of Pursuing the Male Mystique." In those benighted early days, it was not only possible but wholly reasonable to hold out a vision of men and women working together to eradicate the different yet complementary bonds that held each of us back. Soon afterwards, however, things began to change. Dr. Warren Farrell, the leading author for gender equity, became a three-time board member of the New York City NOW and to this day the only man ever to hold such a position. A few years later, Dr. Farrell found himself forced to resign when he realized, to his shock and dismay, that NOW was concerned with promoting special privileges for women. Predictably, not long afterward, NOW removed from its publications the announcement of the award to NCFM and today denies that such an award was ever given.

In August 2004, in Fort Worth, Texas, NCFM held its second face-to-face board meeting in its twenty-seven-year history. NCFM's members are young and old, men and women, conservative and liberal, predominantly straight but also bisexual and gay. Gender equity issues seem to run perpendicular to traditional categories. So it is that, of the two leading equity feminist organizations, the IWF and the WFN, the first is "conservative" and the second is "liberal." As with genital integrity, these labels may be starting to become a bit passé. For me, and for some of my colleagues, the work on genital integrity and gender equity is intimately intertwined. NCFM's vice president is an ardent supporter of genital integrity activism. A position statement drafted by activist organization Not Just Skin appears at the NCFM site.[4] This represents an excellent, but all too rare, example of activists in the two causes making common cause to work toward a shared goal.

3. WHAT IS GENDER EQUITY?

The movement for genuine gender equity is based on the simple premise that males and females should be treated equally in all aspects of life. With a few unfortunate but extremely rare exceptions (roughly equivalent to the small yet still regrettable incidence of anti-Semites working to promote genital integrity), advocates for gender equity are ardent in their support for fair treatment of women. While often disagreeing on tactics and specific priorities, and often diverging strongly in their political orientation, nevertheless, in my experience, most such activists agree on and, indeed, devote their lives to promoting certain goals. They are united in struggling to promote the concept that each of us should be treated fairly and with love and kindness regardless of whether we are male or female. This may not mean a guaranteed equal outcome but it probably means an equal opportunity, level playing field. In a sense, this is also precisely what genital integrity activism is all about.

3.1 Writing About Genital Integrity and Gender Equality

Feminist activists and activists against female genital cutting (FGC), of course, often have drawn connections between their causes, routinely claiming the battle against FGC as a branch of the struggle for women's equality. I agree, just as I believe the fight to protect genital integrity is part of the struggle for gender equity. Male genital integrity has been raised as a feminist issue by Rosemary Romberg,[5] Miriam Pollack,[6] Pamela Bone,[7] and others. A variety of reasons have been mentioned to justify this association: the bonding of mother and child is grossly affected by circumcision, the alliance of women with peace and humanity justifies their opposition to genital alteration, the difficulty of drawing a line between procedures on males and females means that feminists must support genital integrity for everyone, feminists are fighting a power hierarchy that calls for such horrible acts of allegiance as circumcision, and so on.

Parallels between genital integrity and gender equality have been addressed by a number of authors, by some who are primarily identified with the former movement, and by others more closely allied with the latter. From the gender equity side, we have Dr. Warren Farrell,[89] Aaron Kipnis,[10] Jed Diamond,[11] Gordon Clay's monumental MenStuff website,[12] MenWeb,[13] James Whipple in a book produced by NCFM,[14] and others. From the genital integrity activist side, those who have spoken out include Tim Hammond and NOHARMM,[15] Billy Ray Boyd,[16] John Erickson,[17] Joseph Zoske,[18] Sam Keen,[19] and Tom Golden.[20] I published an essay in the gender equity movement's leading magazine examining commonalities between circumcision and men's issues.[21] In another illustration of useful collaboration, the writing team of Frederick Hodges and Jerry Warner published their classic article, "The Right to Our Own Bodies," on one of the leading men's issues websites.[22]

In his article, Zoske makes the trenchant point that we who strive to advance genital integrity don't talk as much as we might about issues such as men's loss and grief and circumcision's impact on a man's overall psychosexual development. Perhaps these subjects are uncomfortable for us, socialized as we are to expect our own and other men's invulnerability or the illusion thereof.

Sam Keen joins the dots together particularly well. Circumcision acts as a "sacrament" in our culture that prepares men for a life in which they are expected to engage in power-based relationships, violence, and warrior mentality:

> Circumcision remains a mythic act whose real significance is stubbornly buried in the unconscious. That men and women who supposedly love their sons refuse to stop this barbaric practice strongly suggests that something powerfully strange is going on here. Feel the violation of your flesh, your being. What indelible message about the meaning of manhood [is] carved into your body? Masculinity requires a wounding of the body, a sacrifice of the natural endowment of sensuality and sexuality. A man is fashioned by a process of subtraction. We gain manhood by the willingness to bear mutilation.[23]

Therapist Golden writes about men remembering their circumcision.[24] Golden uses breathwork, EMDR and other techniques to assist people in moving into old pain to re-experience painful past events and thereby be relieved of the old trauma. (Breathwork is a therapeutic technique using energetic, directed breathing to focus the mind and body.) Golden notes that, for those who do follow this practice, painful memories often turn out to be clustered together. One of the past-life events often re-experienced within a cluster of trauma is the pain of being circumcised. (I can't help noting here that, in a breathwork session, completely unexpectedly, I once very concretely re-experienced my circumcision at infancy.) When Golden first saw this, he was "amazed and shocked" as he hadn't seen circumcision as anything other than a routine procedure. (And so it is for each of us, on each issue to which we have not given previous thought. It took me years to take genital integrity activism seriously.) Golden relates that most of these men felt exactly the same way he initially did. They often were taken aback at the intensity of the pain. Other therapists, Golden reports, have related similar experiences to Golden's.[25]

Golden makes a very direct, concrete link between male circumcision and men's lives. "I am convinced that the practice of circumcision can create trauma in boys. I can't imagine anyone allowing little girls to be cut in their genital area shortly after birth. You would see an avalanche of protest! These boys get through the gauntlet of the birth canal only to find one of their very first experiences with other humans is being cut in one of their most sensitive parts."[26]

Dr. Farrell contributes what may be the most powerful, succinct summary of the intimate association for males of genital integrity and gender equity: "America's reflexive continuation of circumcision-without-research reflects the continuation of our tradition to desensitize boys to feelings of pain, to prepare them not to question the disposability of their bodies any more than they would question the disposability of their foreskins."[27]

4. PARALLELS BETWEEN GENITAL INTEGRITY AND GENDER EQUITY

A number of illuminating parallels link genital integrity and gender equity. Most obviously, some activists participate in both causes and see the two as intimately intertwined. In fact, I believe it is fair to say that most if not all persons who are involved in both movements view them as two sides of the same coin.

4.1 General Parallels

Like genital integrity, gender equity is fundamentally a civil rights movement. Many activists consciously model their thinking and/or their actions on various civil rights struggles from throughout history, most notable the women's suffrage cause of the nineteenth and early twentieth centuries, and the struggle for civil rights for African-Americans, which came into its own in the fifties and sixties.

Another parallel is that both topics seem strange at first blush but paradoxically enough are allied in a very intimate way with basic principles by which most of us agree to regulate our lives and behavior in this post-Nuremburg Trials world. Who would deny that, so far as is possible, discrimination should be rooted out so that all people have the same equal opportunity? One of the few things that might be even more squarely aligned with common sense, especially if you spend any time around babies, is that babies should be loved and safeguarded from all forms of needless harm. However, given how deeply we are entrenched in the prejudices and preconceptions of our world, a moment of epiphany may be required for us to be able to attain the perspective to incorporate these new ideas into our lives.

One relatively unknown, yet well-documented connection between the two issues is that women were barred from participating in Jewish ritual circumcisions in a deliberate effort to minimize women's presence and importance in the temple.[28]

4.2 Legal Parallels

Legal approaches to guaranteeing the right to genital integrity are workable as one prong of a multi-prong strategy to promote gender equity in both directions. (The converse is not as clearly true due to the greater specificity of the genital integrity cause).

Both issues are relatively unfamiliar to the broader legal community, not to mention the general public. Yet, the relationship between the two has been at least implicitly addressed by a number of legal and non-legal scholars.

Both concerns can be rendered easily understandable to the uninitiated through straightforward parallels with more familiar topics. Male genital integrity can be analogized with female genital integrity, and the need for equity for men can be analogized with the need for equity for women.

5. DISTINCTIONS BETWEEN GENITAL INTEGRITY AND GENDER EQUITY

5.1 General Distinctions

Relative to gender equity, genital integrity may in some senses be more immediately palatable to the public. First of all, it may be easier to maintain and fortify a contention that men's equality does not need endorsement due to historical factors. It may be significantly tougher to argue that male baby genitals don't deserve protection just like female baby genitals. However, it certainly cannot be denied that people who are willing to make the attempt do exist! This has been managed to date through a blend of strategies, including lies about an absence of effect on males, explicit and implicit references to male circumcision's alleged religious (primarily Jewish) sacramental role, untruths about medical benefits, misrepresentations about babies' inability to feel pain, and so on.

However, signs are appearing on the horizon that such balancing acts are becoming more difficult to sustain. Recently, a number of influential medical organizations in Australia and New Zealand,[29] British Columbia,[30] and Saskatchewan[31] have forcefully rejected these claims. Regardless of one's views about adults, one tends to see male babies and female babies as both equally innocent and equally vulnerable.

Genital integrity is also likely to strike most people as a less important issue than gender equity. If less important, it may be less necessary to resist. It may also attract a query that I have had countless times about why it merits any attention, but focused repetition of key points should be able to cope with this objection. Finally, the concern about opening the floodgates by admitting the propriety of males' concerns may be less problematic with the more specific issue of genital integrity than it is regarding potentially much broader claims regarding gender equity.

5.2 Legal Distinctions

Genital integrity may be fundamentally more palatable to judges, for similar reasons to why it may be more palatable to the public. Actually, these reasons are not even necessarily separable, since, as mentioned earlier, judges tend to be reluctant to initiate social change, and contrary to popular belief have rarely done so. Again, we are likely to find far lower levels of fear about opening the floodgates.

Genital integrity may temporarily be (or may not be) less legally robust. Currently, courts are quite focused on discrimination, including gender discrimination. These concepts are very familiar to the courts. While it is true that the majority of cases have been decided primarily or exclusively to benefit women, nevertheless, on its terms, the constitution guarantees both men and women the right to freedom from discrimination. Despite the existence of facially discriminatory laws, such as the Violence Against Women Act, whose reconfirmation is apparently being authorized by Congress as this book goes to press, the *theoretical* equality of males and females has never to my knowledge been directly challenged, successfully or otherwise, in a case of record.

The right to genital integrity, by contrast, has not yet been affirmed by the courts despite being contained in a host of human rights documents ratified by the United States and/or applicable to the US through customary law. Courts may view the issue as, at most, an intellectually interesting curiosity, a sideline, possibly an utter waste of time or, at worst, perhaps even an insidious, malevolent attempt to divert attention from truly meritorious issues such as female circumcision.

So, it may not be as big a stretch to decide another case affirming men's right to freedom from sex discrimination, relative to making the first general, on-the-record statement that a widespread practice must stop and genital integrity must be prioritized.

6. LEGAL CASES REGARDING GENITAL INTEGRITY AND GENDER EQUITY

As most activists are aware, no legal cases with precedential value have yet been handed down to help safeguard genital integrity. Individual legal victories have been won, such as the notable settlement achieved in 2003 by plaintiff William Stowell and attorney David Llewellyn.[32]

A few important cases have been decided in state and federal court, affirming men's right to equal treatment. The famous Virginia Military Institute lawsuit held that any law broadly discriminatory against men or women on account of sex is subject to a "strong presumption" that it is unconstitutional, and that such presumption can be rebutted only by an "extremely persuasive justification."[33] Other cases reaffirming men's right to equality — all decided by the United States Supreme Court — include Adkins v. Children's Hospital,[34] Reed v. Reed,[35] Stanley v. Illinois,[36] Frontiero v. Richardson,[37] Taylor v. Louisiana,[38] Weinberger v. Wiesenfeld,[39] Stanton v. Stanton,[40] Craig v. Boren,[41] Califano v. Goldfarb,[42] Duren v. Missouri,[43] Orr v. Orr,[44] Califano v. Wescott,[45] Wengler v. Druggists,[46] Kirchberg v. Feenstra,[47] Mississippi University for Women v. Hogan,[48] and J. E. B. v. Alabama ex rel. T. B.[49] Interestingly, there is some strong suggestion of a parallel between feminism's endorsement of men's equality in the seventies and early eighties and the relatively high number of court cases handed down during that era relative to times before and since then. This striking parallel is not surprising since, contrary to popular belief, courts tend to follow public opinion rather than to lead it. The genuinely

[52] Liptak A. Circumcision opponents use the legal system and legislatures. New York Times;23 January 2003.
[cited 29 August 2004]. URL: http://www.cirp.org/news/ newyorktimes01-23-03/

[53] Medina G. Circuncision ilegal de recien nacido hispano. El Diario/La Prensa 4 Aug 2003:1.

[54] Hageman *et al.,* v. Stanek *et al.,* Minnesota Supreme Court No. A03-2045.

[55] Booth v. Hvass, 302 F. 3d 849 (8th Cir. 2002), cert. den., 537 U.S. 1108 (January 13, 2003).

[56] California District Court Case Number 05CS01530 (Sacramento).

[57] MSNBC. Donahue show. 18 December 2002.

[58] CNN. California recall election [segment regarding Warren Farrell and gender equity movement]. 27 August 2003.

[59] NBC. Man ordered to support child he says isn't his. 2 March 2004. [cited 30 August 2004]. URL:www.nbc4.tv.

[60] Fox6 News. 6 August 2005. [cited 26 October 2005]. URL: www.ncfmla.org/media_coverage.html.

[61] Angelucci M. Fairness behind bars. Los Angeles Times 25 June 2005;B21. [cited 26 October 2005]. URL: www.latimes.com.

[62] Angelucci M. Men are victims, too. California State Bar Journal October 2005:8. [cited 26 October 2005]. URL: http://calbar.ca.gov/state/calbar/calbar_cbj.jsp?

[63] McRae S. Los Angeles Daily Journal 3 May 2004;1.

[64] Showtime. Penn & Teller: Bullshit! 25 April 2005.

[65] Chester-Taxin S. Penis page. Penthouse November 2001;120.

[66] Newman B. 'Intactivists' seek to undo a long-practiced ritual. Wall Street Journal 28 December 2000;A1.

[67] Hess M. Child advocates urge NYC to regulate circumcision. 30 September 2005. URL: http://news.yahoo.com/news?tmpl=story&u=/prweb/20050930/bs_prweb/prweb291936_1

[68] McElroy W. Routine procedure or child abuse? 6 August 2002. [cited 30 August 2004]. URL: www.fox.com and www.ifeminists.com.

[69] Ollivier DS. Circumcision in America. Salon 26 October 1998 and 27 October 1998. [cited 26 October 2005]. URL: http://www.salon.com/mwt/feature/1998/10/26feature.html.

[70] Zeller S. Activists fight circumcision. National Journal 29 June 2002;1962.

[71] Roig-Franzia M. North Carolina group fights state funding for circumcisions. Washington Post 23 June 2002;A2.

[72] Rice L. The unkindest cut. Harvard Law Bulletin Fall 2002;55. [cited 30 August 2004]. URL: http://www.law.harvard.edu/alumni/bulletin/2002/fall/classnotes_13.html.

[73] Fernandez L. Saying no to circumcision: a few Jews are breaking with tradition and having alternative birth rites instead. San Jose Mercury News 12 July 2003;F1.

Chapter 15

INCREASING AWARENESS OF IATROGENIC DAMAGE CONSEQUENT TO MALE CIRCUMCISION

Jim Bigelow and R. Wayne Griffiths
National Organization of Restoring Men (NORM)

Abstract: An initial attitude/awareness survey was conducted between 1991 and 1994. Subjects were drawn from a population of circumcised males known to favor intactness. Results were presented August 1998 at the Fifth International Symposium on Sexual Mutilations at Oxford, England. Subjects were self-selected based on (1) their seeking foreskin restoration information and (2) their decision to fill out and return the questionnaire provided. In October 2003, an Internet version of the questionnaire was presented to the general public. The two self-selection factors in the current survey are: (1) men who choose to visit www.norm.org and (2) those who choose to respond. Data collection (N = 1334) for this paper closed 31 March 2004. Respondents were asked: (1) for demographic information, (2) to indicate their circumcision status, (3) to indicate various attitudes toward that status, and (4) to report their knowledge of or participation in a foreskin restoration program. The results show that most men who reported negative attitudes toward being circumcised were circumcised in infancy/childhood. It was further found that men who where circumcised as adolescents or adults had more positive attitudes toward their circumcision and reported significantly higher self-esteem as compared to those who were circumcised earlier in life. Finally, it was found that one of the more salient factors related to success in a restoration program was partner support.

1. INTRODUCTION

At the Fifth International Symposium on Sexual Mutilations, held 5-7 August 1998, at Oxford, England, the results of a survey of two-hundred-forty self-selected circumcised males were presented. The results indicated

G.C. Denniston et al. (eds.), Bodily Integrity and the Politics of Circumcision, 165–175.

that circumcised men in the United States were becoming increasingly aware of the negative effects of routine circumcision. (See, *Male and Female Circumcision: Medical, Legal, and Ethical Considerations in Pediatric Practice*: 1999 Kluwer Academic/Plenum Press).

That original (1991-1994) survey was done by sending a printed questionnaire with a return envelope to men who sought foreskin restoration information. The subject population was limited to those men who, having identified themselves as dissatisfied with their circumcision status, felt strongly enough to respond to the questionnaire.

The advent of the Internet, with its impact of unprecedented access to information and interpersonal contact, has resulted in possibilities never before imagined. As a result, this present survey is available to men across a broad spectrum — from men who have never questioned male circumcision, much less thought about the possibility of foreskin restoration, to those seeking specific restoration information. While some subjects sought out the website for the National Organization of Restoring Men (NORM), others simply stumbled upon it. More than half of all the respondents reported that they first learned of restoration via the Internet. One consequence of the broader subject population is that some of the respondents were pleased to be circumcised. This fact alone has expanded the scope of the inquiry.

A questionnaire similar to the 1991-1994 survey reported above is posted on the NORM website www.norm.org. The volume of participation has been overwhelming. The questionnaire was first posted in October 2003 and, as of 31 March 2004, the participation was cut off in order to prepare for this symposium, N = 1334. Duplicates were identified by their "remote name," age, and identical information and were eliminated. There are continuing responses to the questionnaire, approximately ten per day. At the time of this presentation, 2300 responses had been collected. In 2005, it is proposed that the total group be looked at to discover if more precise response patterns would emerge. It is further hoped that this larger subject population will help in restructuring the questionnaire so that, in the future, more precise, definitive questions can be studied.

The population that completed the questionnaire is most likely not representative of the circumcised male population at large. Self-selected subject populations always have an element of bias. As a result, any extrapolation from the results of this survey to the general circumcised population must be done with care. Examples of elements of recognized bias include subjects (1) who have access to a computer, (2) who know how to navigate the Internet, and (3) who are sufficiently interested in both circumcision and restoration to consider the related subjects and respond to the survey. In addition, a subject must be willing to share his own feelings and experience. These data do, however, clearly represent a growing subpopulation of circumcised males who are aware of their circumcision status and are conscious of their attitudes toward it.

2. METHODOLOGY

After several years of correspondence and contact with men seeking restoration information (including conducting the original 1991-1994 survey), it became apparent that these men voiced similar attitudes and had complaints in common. It was decided to construct for the Internet what would obviously be a preliminary questionnaire (Table 1). The most basic purpose was to discover whether or not men would reveal very private information about their circumcision status and/or expose their personal feelings to an organization they typically knew little or nothing about. A second purpose was to discover if attitude patterns could be documented and if certain aspects of the individual's circumcision experience might relate in particular ways to such attitudes. In most instances, the individual's circumcision was done without the benefit of personal choice. It was hoped that this survey would codify the negative attitudes, anger, and often the despair, which were regularly voiced, to such an extent that parents and the medical profession would take note.

Table 1. NORM Survey on Circumcision and Foreskin Restoration

Please complete all the questions that apply to you. All responses are anonymous.

1. How old are you?
(Enter a two digit number, for example "35")

2. At what age were you circumcised?
(Please choose one)
Infant (less than 1 year)
Child (1-10 years)
Adolescent (11-17 years)
Adult (18 or older)

3. Are you happy you are circumcised?
Yes No Neutral - Don't care one way or the other

4. How has being circumcised affected your sex life?
Positive effect on sex
Has improved sex
More sensation
Neutral
No effect on my sex life
Negative effect on sex and/or loss of sensation

5. **Has being circumcised made you feel?**
   (Select one on each line)
   Superior Inferior Neither Superior or Inferior
   More Masculine Less Masculine Neither More nor Less Masculine
   More Attractive Less Attractive Neither More nor Less Attractive
   Admired Object of Ridicule Neither Ridiculed nor Admired
   More Confident Less Confident No Effect on Confidence
   Proud Embarrassed Neither Proud nor Embarrassed
6. **When you discovered there were males who were not circumcised, what feelings arose?**
   (Check all that apply)
   Envy Anger and/or rage
   Curiosity Loss, Feelings of sadness
   Admiration Neutral, no reaction
7. **Are you envious or jealous of men who are intact (not circumcised)?**
   Yes No
8. **When did you find out about foreskin restoration?**
   Please choose a time period (2000-2003) (1995-1999) (1990-1994) (Before 1989)
9. **Where did you first hear about foreskin restoration?**
   NORM (website, phone, e-mail)
   Foreskin Restoration Website (other than NORM, NOCIRC, or NOHARM)
   Book (such as *The Joy of Uncircumcising!*)
   Friend or Relative, Web
   Newspaper, Magazine Article, TV Show, Information Booth (At a fair or conference, for example)
   Other Source, Radio, NOHARMM or NOCIRC website
   Literature
10. **What did you do when you found out about foreskin restoration?**
    (Please choose one)
    No Action,
    Looked on the Internet for information
    Contacted NORM or other group
    Talked to a doctor
    Talked to a counselor (psychologist, minister, etc.)
    Looked for more print information (books, leaflets, etc.)
    Started restoring with my own improvised technique

If you have begun restoring your foreskin, please answer the following questions:

11. How long have you been restoring your foreskin?
Less than 6 months
6 months to 1 year
1-2 years
2-3 years
More than 3 years

12. What results have you had so far? (Skin growth and/or increased sensitivity)
(Please choose one)
No results yet
Poor or minimal results
Good results, average progress
Great results, better than average
Exceptional results, better than ever imagined

13. How do you feel after regaining your foreskin (or attempting to)?
(Check one on each line)
Natural/Whole Freakish Neither
Satisfied Dissatisfied Neither
More Attractive Less Attractive Neither More or Less Attractive
Object of Envy Object of Ridicule Neither Ridiculed or Admired
Self Assured Self Conscious Neither

14. Have you consulted a physician, doctor, or psychologist regarding your restoration?
Yes
No

If "Yes," please answer 14b, 14c, and 14d

14a. What type of health professional did you consult?
(Please choose one)
MD - General Practice
MD – Urologist, DO
Psychologist, Counselor, Psychiatrist, Sexologist
Plastic Surgeon
Alternative Medicine (ND, for example)
Other health professional

14b. How did the doctor respond?
Sensitively, Compassionately, Interested, and Supportive
Indifferent, Didn't care
Ridicule, Suggested it was a psychological issue

14c. Was the doctor aware of foreskin restoration before you mentioned it?

Yes
No

15. If you have a spouse or partner, what are your partner's feelings/reactions about your interest/participation in a restoration regimen/program?

(Check one)
Very supportive, excited, encouraging, suggested restoration
Supportive, curious
Indifferent, neutral, didn't care one way or the other
Unsupportive, disdain, preferred you didn't restore
Very negative, ridicule, put downs, threatened you if you restored

The following questions are optional and are for demographic and background information.

What race or geographic region do you most identify with?

(Choose one)
White (American or non-European born)
Black (non-African born)
Asian
European (western)
European (eastern and former Soviet countries)
African
South or Central American
Middle Eastern
Native American (North or South America)

In what state or country (if born outside the US). were you circumcised?

AB AK AL AR AZ BC
CA CO CT DC DE FL GA
HI IA ID IL IN KS KY LA
MA MB MD ME MI MN MO MS MT
NB NC ND NE NH NJ NL NM NS NT NU NV NY
OH OK ON OR PA PE QC RI
SC SD SK TN TX UT VA VT
WA WI WV WY YT
Or, if a country other than the US, please write your answer below:

Of the major world religions, which of the following most closely matches your family's religious background or affiliation?

Atheist
Christian
Catholic
Jewish
Buddhist
Hindu
Sikh
Confucian
Bahia
Jain
Shinto
Humanist
Christian Scientist
The Church of Jesus Christ of Latter Day Saints
Unitarian
Universalist
None of the above

Did your family's religion play a role in your being circumcised?

Yes
No
Not sure or doesn't apply

Finally, you may submit any further information, comments, or clarifications to the questions in this survey that you would like to provide. If you have specific questions you need answered or if you want more information, please e-mail questions@norm.org.

If you would like to tell your story regarding circumcision and/or restoration, either for posting on the NORM website or for an upcoming book on men's (and their partner's) experiences of foreskin restoration, please send it to story@norm.org.

Results were tallied and, where appropriate, additional statistical analysis was done to better understand the composition of this particular sub-population of males and to document and report the findings. Factors that were considered included: (1) age of respondent, (2) age at which the individual was circumcised, (3) general description of the individual's attitude toward being circumcised, (4) racial/regional identification, and (5) religious affiliation.

In addition to the questions of general attitude about being circumcised, the subjects were asked a range of questions regarding specific feelings related to their circumcision. Did the individual's circumcision cause him to feel superior/inferior, more confident/less confident, etc., as well as specific questions regarding the perceived affects of circumcision on his sex life.

Finally, respondents were asked to indicate their knowledge of and/or participation in foreskin restoration. Among the questions asked were indicators such as: (1) were they involved in restoration, (2) length/duration of their program(s), (3) success or frustration with their attempt(s) at restoration, and (4) if in an ongoing relationship/partnership, was their partner supportive of their desire and/or regimen to restore their foreskin.

3. RESULTS (INTERNET QUESTIONNAIRE N = 1334)

Personal descriptive data

1. AGE

14%-15%	10-20 yrs old
65%	20-50 yrs old
20%	50-70 yrs old

2. AGE WHEN CIRCUMCISED

84% -85%	as infants
7%	as children under 10 years of age
3%	as teenagers
5%	as adults

3. GENERAL ATTITUDE TOWARD BEING CIRCUMCISED

84%	not happy
10%	neutral
6%	happy

4. RACIAL/REGIONAL IDENTIFICATION (when given)

2%	Asian
1.6%	Black (African-American)
2.2%	Native American
73.5%	White American

5. RELIGIOUS AFFILIATION

82%	Christian
3.4%	Jewish
5%	Atheists
9.6%	Affiliated with other philosophies or none of the above.

4. COMPARATIVE FINDINGS

1. The sheer volume of response to the current survey provides the answer to the first question. Hundreds of men seem more than willing, even anxious, to reveal both their circumcision status and to document their attitudes and feelings related to that status.

2. While no specific statistical analysis of this factor was performed, the vast majority of subjects who reported a negative attitude toward being circumcised were circumcised in infancy/childhood.

3. Those subjects who were circumcised as either adolescents or adults tended to express more positive feelings about their circumcision and about themselves relative to circumcision.

Subjects were provided six scales and asked to rate how their circumcision made them feel. These scales included: superior/inferior, more masculine/less masculine, more attractive/less attractive, admired/ridiculed, more confident/less confident, and proud/embarrassed. Scores were assigned to responses: 1 for negative responses, +1 for positive responses, and 0 for neutral responses. Taken together, positive scores on these scales were taken to indicate higher self-esteem and negative scores were taken to indicate lower self-esteem.

There was no significant difference noted on any of these scales between those circumcised as adults or as adolescents. However, when those who were circumcised as adolescents or adults were compared to those circumcised as infants or children, those circumcised later in life had higher scores on five of the scales: superiority, masculinity, attractiveness, confidence, and pride. All these differences are significant ($p < .05$).

4. Several cross comparisons were made between the age at which circumcision was performed and indicators of self-esteem. When the scores for various aspects of self-esteem were combined, the results indicated that those who were circumcised as adolescents or adults had significantly higher scores for self-esteem than either those circumcised as infants or children. All differences were statistically significant ($p < .05$).

5. In addition to the more obvious factors contributing to an individual's success at restoration, the fact of partner support was found to be of significant importance ($p = .0222$).

5. DISCUSSION

While the medical profession typically maintains that circumcised males are either neutral or positive toward their circumcision, the responses to the Internet questionnaire would suggest that an increasing number of circumcised males resent the medical procedure that was imposed upon them.

It has long been speculated that men who have been circumcised as infants harbor a sense of victimization. The fact that the majority of men who reported resentment of their circumcision were circumcised as infants supports this notion. It seems apparent that not being included in the decision to be circumcised is a major contributor to such negative feelings.

On the other hand, those circumcised as adolescents or adults tend to have more positive feelings about their circumcision. These findings certainly suggest that the circumcision decision should be left to the individual when he is old enough to at least enter into the decision process. Furthermore, in our society it might need to be pointed out that adult males are not only capable of making the circumcision decision for themselves but tend to be more frequently happy with that decision if they have made it. In addition, adult males who were at least partially involved in their own circumcision decision tend to report a higher level of self-esteem.

The purpose of NORM is to support men in their efforts at foreskin restoration. It is, therefore, important to know what factors contribute to an individual's decision to seek to restore his foreskin. A major factor, age at circumcision, is a variable over which the potential restorer most frequently has had no control. It would seem, however, that a lifelong resentment of one's early circumcision is a major motivating factor to sustain restoration efforts.

If, in most cases, circumcised males report decreased self-esteem, the restoration process is an obvious means by which such lowered self-esteem can be enhanced. Given the social taboos regarding the penis and the foreskin, it is not surprising that partner support was found to be a significant factor in the successful completion of a restoration program. In the NORM support group meetings, which began in February 1990, the subject of partner reaction, attitude, and support has often been the focus of discussion. While restoration is an infinitely private matter, for those who share their lives and bodies in an intimate partnership, the value of positive support from one's partner cannot be overestimated.

6. CONCLUSION

More research needs to be done both (1) in an effort to help those men who already feel mutilated by circumcision find some sense of resolution and hope and (2) to find additional factors to support the end of routine infant male circumcision so that in the future no male will be circumcised against his will.

ACKNOWLEDGMENTS

James Schinnerer, Member of the Board of Directors, NORM, for maintaining the web site, creating the Internet survey, and collecting the data. Robert Van Howe, MD, for statistical analysis.

Chapter 16

A SURVEY OF SUBJECTIVE FORESKIN SENSATION IN 600 INTACT MEN

Peter J. Ball
NORM-UK

Abstract: The foreskin contributes immensely to the enjoyment of sex. Phimosis may mar this enjoyment but is a condition that may well be preventable by appropriate early instruction. The foreskin is a sexual organ in its own right, which is ablated forever by the act of circumcision.

Most people are aware that circumcision removes the foreskin and permanently exposes the glans of the penis. This results in the thickening or keratinization of the glans. For some, this is a bonus because it enables them to delay an orgasm, for others it is a curse, making efforts to reach orgasm tedious or even impossible. I believe that many women and most circumcised men are totally unaware of the role the foreskin plays in the enjoyment of sex. To increase this awareness, I have asked more than six-hundred intact men to answer twenty questions related to their subjective experiences of foreskin sensation.

The subjects were drawn from a variety of sources: men whose intact status was already known to me; referrals from these men, many of whom replied to an invitation to complete my questionnaire, which was placed on the NORM-UK website and on a naturist website; and some were recruited from subscribers to three Yahoo foreskin-related internet discussion groups. A large group came from men whose foreskins were too tight and who had sought advice via the NORM-UK website. During the preparation of my questionnaire, I showed it to Dr. John Taylor, who suggested some extra questions. I shall discuss the questions posed and the replies received.

G.C. Denniston et al. (eds.), Bodily Integrity and the Politics of Circumcision, 177—188.

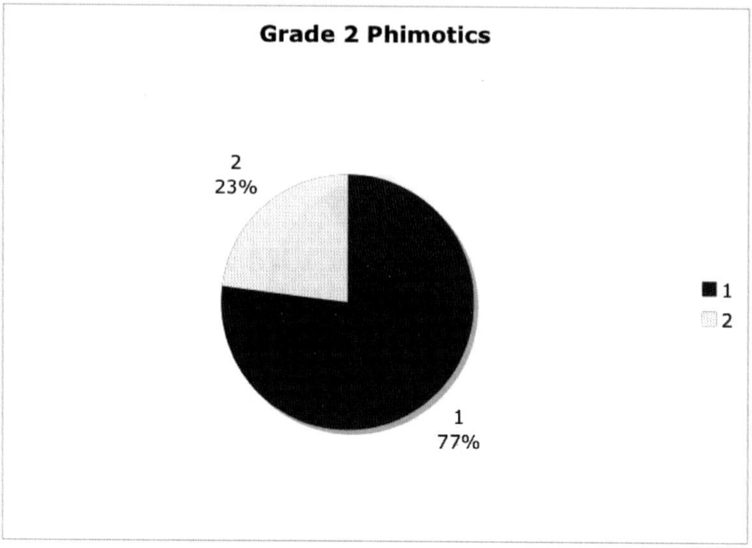

Figure 16-3. Grade 2 Phimotics.

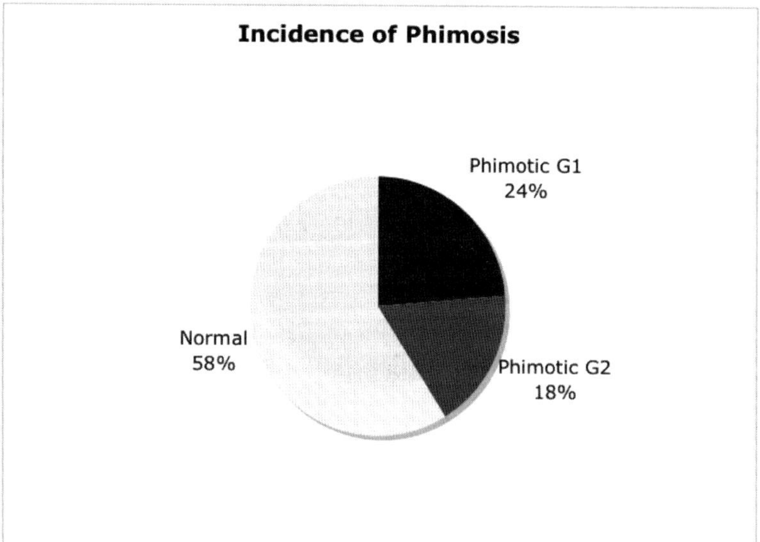

Figure 16-4. Incidence of Phimosis.

The figures do not reflect the natural distribution of phimosis, which is abnormally high in this study because of the addition of all my phimotic subjects. About ten percent of self-selected subjects were phimotic by my definition.

QUESTION 4. "WHAT DO YOU FEEL AS THE FORESKIN IS RETRACTED?"

QUESTION 5. "IS IT PLEASURABLE?"

The answers to these questions reveal a significant difference in experience between the normal and the phimotic. Pleasurable feelings were reported by eighty percent of the normal, while only three percent of the phimotics had any pleasant feelings, and they were all reported by the Grade1 group. The reason for the difference is clear. In the phimotic group, the early onset of tightness or pain negated any pleasure that they might have experienced from stimulation of the ridged band.

We are indebted to Dr. John Taylor, *et al.,* for demonstrating the nature of the human foreskin in microscopic detail. The description of the "ridged band" gave an anatomical basis for the pleasure that the foreskin provides during sexual activity. In 1996, Dr. John Taylor and colleagues wrote a paper in which they describe the histological appearance of an area 2-3 mm within the mucocutaneous junction of the foreskin that was packed with end organs unique to that area, which were sensitive to touch, movement, and stretching. This region can be seen with the naked eye as a series of ridges in the inner foreskin that encircle the tip and converge on the underside of the glans into the frenulum. Dr. Taylor called this the *ridged band*. Details of his discovery can be found at http://research.cirp.org.

QUESTION 6. "IS THE FEELING RELATED TO ANY PARTICULAR PART OF YOUR FORESKIN AND IF SO WHERE?"

The number of men reporting pleasure was 90% in the normal group and 46% in Grade 1 phimotics and 24% in the Grade 2 group. The reason for the low figures in the phimotic groups was that pain frequently replaced pleasure as the foreskin was retracted. The descriptions from the normal group were rich and varied, for example, "Extremely pleasurable," "Very horny sensation," "All sensations are more intense," "Yes it's sensational," "A very erotic sensation." Some 60% of normal foreskin owners related the site as being within the tip of the foreskin. Some of the phimotics could relate the feelings to the end of the foreskin.

QUESTION 7. "IS THIS FEELING DIFFERENT IF THE FORESKIN IS RETRACTING TO ACCOMMODATE AN ERECTION?"

A few of the normal group said that the erogenous feelings were enhanced, while the phimotics all described pain.

QUESTION 8. "CAN YOU DESCRIBE IT AND SAY WHERE IT IS FELT?"

QUESTION 9. "CAN YOU DISTINGUISH BETWEEN SENSATIONS FROM THE GLANS AND FROM THE FORESKIN?"

Some sixty percent thought they could distinguish feelings between the foreskin and the glans.

QUESTION 10. "CAN YOU DESCRIBE WHAT FEELINGS THE FRENULUM PRODUCES AS THE FORESKIN IS RETRACTED"

Turning now to the frenulum, this has been variously described as the g-spot, the love chord, the anchor. This is indisputably the most erogenous zone of the penis.

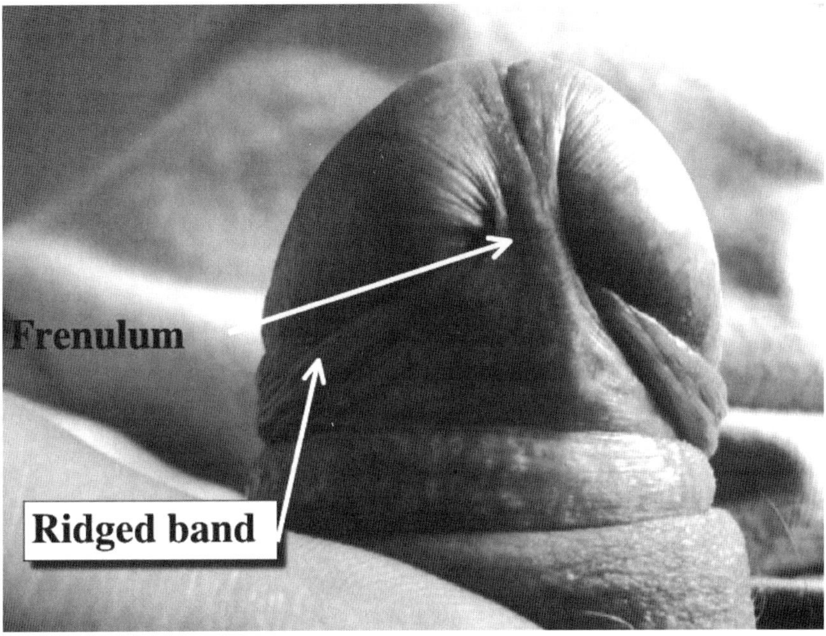

Figure 16-5. Frenulum

QUESTION 11. "HOW DOES THIS DIFFER WHEN IT IS STRETCHED DURING AN ERECTION?"

These questions brought into focus another variability of the intact penis, the length and elasticity of the frenulum. Many frenula are long enough to allow the foreskin to be retracted right down to the base of the shaft. Some become painful the moment the foreskin is moved across the glans. If it is short or inelastic, its owner usually describes it as tight. The owner will describe the frenulum tugging the glans ventrally when the foreskin is retracted. Sometimes this caused no problem but, for many, particularly in the phimotic group, pain on stretching negated any erotic feelings.

In the normal group, sixty-five percent found it highly pleasurable to be stimulated, and this increased when the frenulum was stretched. Some men reported that they could reach orgasm by stimulating this area alone.

QUESTIONS 12 AND 13. "HAVE YOU HEARD OF TAYLOR'S RIDGED BAND? IF SO, HAS IT ALTERED YOUR ABILITY TO LOCALIZE SENSATION?"

The subjects were also asked whether they had heard of Taylor's ridged band and, if so, had it altered their ability to localize sensation? Only one percent were aware of the ridged band and of those about half said it had improved their ability.

QUESTION 14. "CAN YOU IDENTIFY ANY PARTICULAR PART OF YOUR PENIS OR MOVEMENT THAT TRIGGERS EJACULATION?"

This question produced a variety of answers. Those in the intact group mostly reported that to and fro movement of the foreskin over the glans was the trigger. A few said it was stimulation of the glans and not sensation from the foreskin that triggered the climax. Some said stimulation of the frenulum and others said that kneading the tip of the foreskin without retraction could cause ejaculation. The phimotic group reported that movement of the foreskin around the glans was the trigger.

QUESTION 15. "ANY COMMENTS ON FEELINGS RELATED TO THE FORESKIN DURING INTERCOURSE OR MASTURBATION WOULD BE WELCOME."

This produced the biggest variety of descriptions. Here, Dr. Taylor's description of the penis rolling in and out of its skin tube during intercourse was echoed by the answers to this question. Some fifty percent of intact men intimated here that, for them, the foreskin was essential for the enjoyment of sex whether for masturbation or intercourse. A few tried wearing the foreskin in the fully retracted position for a few days in an attempt to experience what it must feel like to be circumcised. One man kept his foreskin retracted for over a year. All were very relieved to return the foreskin to its normal position. Many took this opportunity to express their thankfulness that they had been left intact and to deplore the act of circumcision. I referred to Kirsten O'Hara's book *Sex As Nature Intended It.*

She wrote this book for the American male, most of whom had been circumcised. She describes in detail her experiences with intact and circumcised men and explains why she believes that intercourse with intact men is infinitely more pleasurable than with circumcised.

QUESTION 16 "WOULD YOU LIKE TO BE INFORMED OF THE RESULTS OF THIS SURVEY?"

Ninety-nine percent said "Yes."

QUESTION 17. "WHAT INFORMATION WERE YOU GIVEN AS A CHILD ABOUT THE CARE OF YOUR FORESKIN?"

I thought there could be a connection between poor parental communication and subsequent phimosis. This is demonstrated by the pie chart shown below:

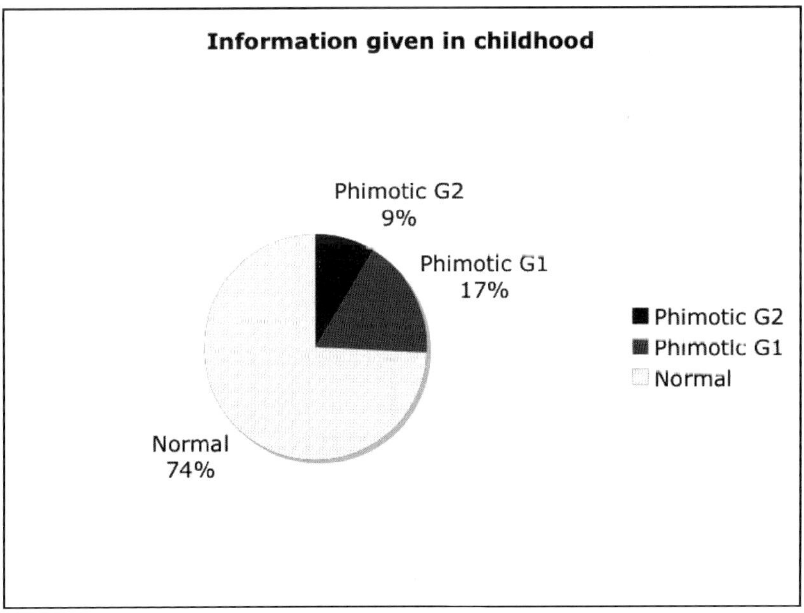

Figure 16-6. Information Given In Childhood.

In the phimotic group, the figure was twenty-six percent. Seventeen percent of Group 1 phimotics (the milder condition) and nine percent of Group 2 (the more severe phimotic) had been told by the parent or caregiver to tell the child that the foreskin was a retractable structure and that it should be washed regularly when any retraction was possible. I think that it is not every little boy that dares to pull back his foreskin, particularly if it hurts or feels uncomfortable.

He needs reassurance that it is safe to do so. I found the degree of ignorance among the phimotics quite astonishing. Many phimotics had reached puberty totally unaware that the foreskin should be able to retract and had never seen a circumcised penis. With family units getting smaller and diminishing school sports facilities involving communal changing rooms, the opportunities for genital comparison are limited. I expected to find that phimosis occurred because of lack of this information. I think these figures confirm this correlation.

When the boy gets older, he should be encouraged to attempt to retract his foreskin during his bath, and that, once retraction is achieved, he should be told that it is normal hygiene to wash the glans and foreskin with warm water daily. Soap can be a foreskin irritant and should be avoided. Remember the possibility of a newly stretched foreskin or prematurely retracted foreskin slipping behind the glans for the first time and the boy having difficulty in returning it to its normal position over the glans. This condition is called paraphimosis and is easily corrected by gently squeezing accumulated blood out of the glans and gently but firmly "popping" the foreskin back into place.

If the parent fails to return the foreskin to its normal position, this can become a medical emergency that requires immediate attention. It does *not* require circumcision. Obstetricians, pediatricians, urologists, general practitioners, health visitors, nurses, and parents are all potential threats to the child's foreskin. He will allow retraction to occur at a pace appropriate for him if he has been appropriately instructed and encouraged.

QUESTION 18. "HOW OLD WERE YOU WHEN YOU COULD FIRST FULLY RETRACT YOUR FORESKIN?"

About sixty percent could remember the time of first retraction. The answers ranged from "Always to Never." This is a graph showing the age incidence.

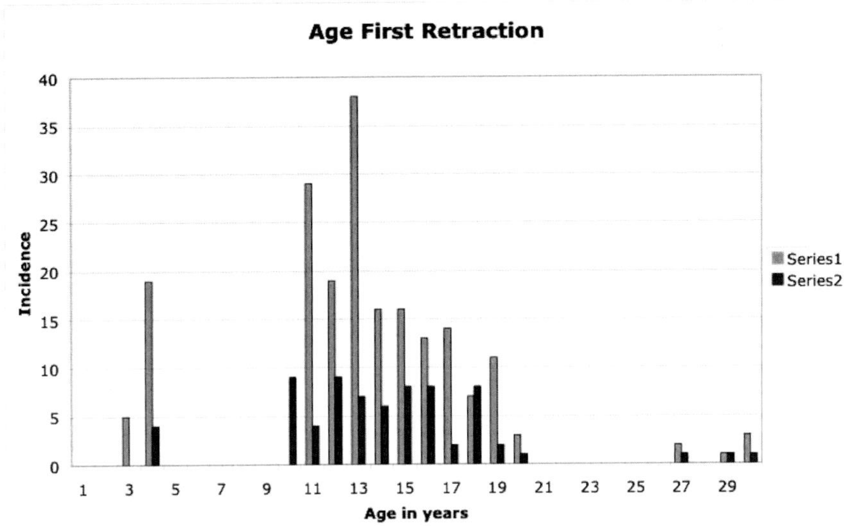

Figure 16-7. Age of First Retraction.

The light shafts are the Normal and the dark the Phimotic. There were peaks of incidence at three and then ten and twelve years in the normal group. The phimotic groups had a wide range of ages of first retraction. I was looking for a correlation between age of first retraction and phimosis in later life. I suspected that boys who had retracted their foreskins before puberty were unlikely to suffer from phimosis in later life. This is confirmed by this survey. Only five percent reported phimosis of late onset (older than thirty), some of whom were classified as suffering from BXO (balanitis xerotica obliterans). BXO is a skin disease, sometimes called lichen sclerosis. It can attack all ages but is more commonly seen in older men. The tip of the foreskin becomes white, scarred, and unretractable. It can involve the glans and can cause urethral strictures. It is treatable either by strong steroid creams or circumcision.

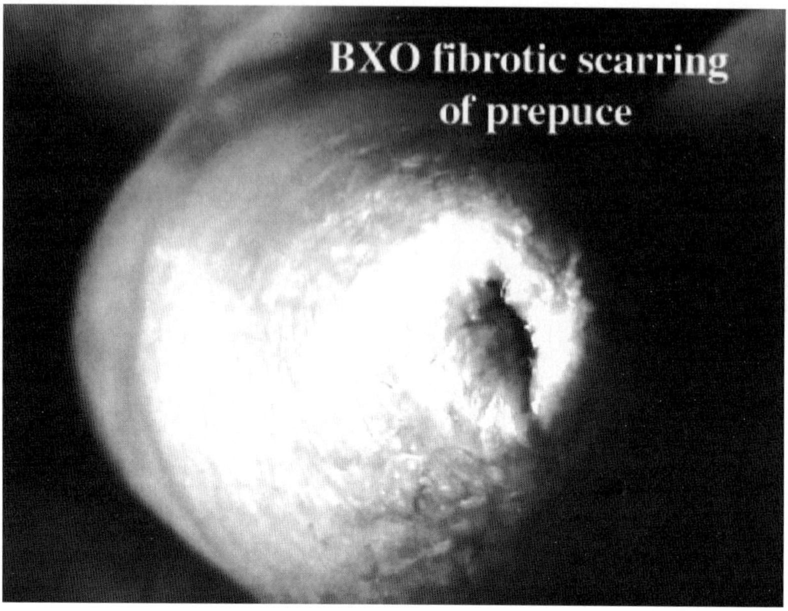

Figure 16-8. BXO.

QUESTION 19. "HAS YOUR FATHER OR ANY CLOSE FAMILY RELATIVE BEEN CIRCUMCISED?"

I suspected that fathers who had been circumcised were less likely to give useful information about the care of their son's foreskin since they were less likely to know what to tell them. The figures were inconclusive.

My thanks to the six hundred men who pondered so deeply to answer my questions.

REFERENCES

1. Taylor JR, Lockwood AP, Taylor AJ. The prepuce: specialized mucosa of the penis and its loss to circumcision. Br J Urol. 1996;77:291-5.
2. http://research.cirp.org.

Chapter 17

HUMAN RIGHTS ADVANCES IN THE UNITED STATES

George C. Denniston
Doctors Opposing Circumcision, Seattle Washington

1. INTRODUCTION

According to legend, when English King Richard I (1157-1199), affectionately known as "Richard the Lion Hearted," rode off to the Holy Land to fight in the Third Crusade, Robin Hood was among those who remained loyal to the absent King. When Richard's brother John (1166-1216) ascended the vacated throne in 1199, he laid on such heavy taxation that the feudal barons rebelled. Meeting King John at Runnymede in 1215, the barons made him sign the Magna Carta. This was the famous first step in limiting the long-standing divine right of kings. Led by Earl Simon de Montford, the English nobles arranged the first meeting of Parliament in 1265, and the rights of the people in relation to the King were further strengthened.

The American Constitution has been the brightest light for human rights since the Magna Carta. The history of the United States has been one of uneven progress toward more freedom and toward fewer human rights violations. Since its inception, the human rights record of the United States was tarnished by the legal existence and persistence of several serious human rights violations: slavery; the denial of female suffrage; and racial segregation. Initially, governmental authorities denied that these institutions were abusive. Later, as a result of grass roots efforts, these abuses were finally acknowledged as such and abolished. Today, it is increasingly

G.C. Denniston et al. (eds.), Bodily Integrity and the Politics of Circumcision, 189—201.

acknowledged that all humans have a right to an intact body – to keep all of the body parts with which they were born. Furthermore, there is a growing awareness that this fundamental right is violated on a grand scale by the medical industry. Progressive and enlightened forces are now seeking to abolish this violation and restore to everyone the right to remain physically intact.

2. SLAVERY

Slavery, the condition in which a person is held as property of another person, existed before the dawn of history, probably evolving out of the enforced servitude of prisoners of war. In the latter part of the fifteenth century, the discovery of the New World by Christopher Columbus provided a new impetus for slavery. The cultivation of tropical crops, such as sugar cane, tobacco, coffee, and cotton, was labor intensive. Soon, merchants from Spain, Portugal, Britain, France, and Holland were transporting African slaves to the New World to meet the labor demands of New World colonies. It is estimated that a total of ten million slaves were transported to the Americas between 1450 and 1810. The trade peaked shortly before its demise. Four million slaves came to the Americas between 1760 and 1810, the fifty-year period just before the importation of new African slaves was outlawed in 1810.

The Enlightenment period of the eighteenth century led to the widespread recognition of human individuality and dignity. Each human is unique and, as the philosopher Immanuel Kant stated, "A thing has a price if any substitute or equivalent can be found for it. It has dignity or worthiness if it admits no equivalent." The English began to think differently about human beings and slavery and, in 1833, slavery was abolished throughout the English colonies.

In the United States, however, no such consensus could be reached, and the morality of slavery remained a topic of debate. In 1776, the Declaration of Independence almost failed to pass because of the slavery issue. Southern states refused to sign unless the divisive issue of slavery was ignored. Benjamin Franklin gave a stirring speech in which he appealed to the states first to consider the enemy, Britain, and join together against that common enemy. He left no doubt in anyone's mind that the problem of slavery would continue to fester and plague the new nation.

By 1850, slavery had become the most important issue in American politics. The South believed it to be proper and necessary. The North thought slavery was abominable and attributed the backwardness of the South to it. William Lloyd Garrison referred to it as "the peculiar institution of human bondage." In 1852, Harriet Beecher Stowe wrote the novel, *Uncle Tom's Cabin*, which exposed the brutalities of slavery.

Another concern of the growing nation was whether slavery should spread into the new western territories. In 1820, the Missouri Compromise attempted to settle the question by balancing the number of free and slave states and allowing Missouri to enter the union as a slave state. The Compromise also dictated, in part, that no territories above 36° 30' latitude could enter the union as slave states. Two decades later, in 1846, a slave in St. Louis named Dred Scott challenged the Compromise by bringing suit for his freedom. The case reached the Supreme Court in 1856 and, in 1857, the Court ruled against Scott and declared the Missouri Compromise unconstitutional.

Both political parties, the Democrats and the Whigs, were badly split over the issue. Mutual hatred was whipped up by inflammatory speeches. The Whig Party folded over their internal divisions on this issue, and the Republican Party was formed, promptly nominating Abraham Lincoln. This new party, born out of the struggle against slavery, enjoyed unprecedented success. From 1860 until 1912, when Woodrow Wilson was elected to the White House, a period of fifty-two years, Republicans, with only one exception (Grover Cleveland), held the Presidency of the United States.

3. WOMEN'S SUFFRAGE

Throughout the nineteenth century, many prominent women publicly objected to the fact that women were forbidden by law to participate in the governing process. They demanded the right to vote. Two years after black males won the right to vote in 1870, women tried to vote in an election and failed. They organized, learned much from the abolitionists, and pressed forward. In 1869, Wyoming Territory gave women the vote. In 1879, the Supreme Court permitted women lawyers to argue before the Court. In 1896, Idaho women gained suffrage by amending their state constitution. In 1905, Emmeline Parkhurst of England resorted to arson, bombings, hunger strikes, and window smashing to focus attention on women's suffrage. In 1910, the women of Washington State obtained the vote by a constitutional amendment. In 1917, after picketing the White House, President Wilson told women that he endorsed equal suffrage. Finally, in 1919, an amendment to the US Constitution was adopted by joint resolution of Congress, and sent to the States for ratification. On 26 August 1920, Tennessee's vote in favor of female suffrage provided the required two-thirds majority, and the 19th Amendment to the Constitution was made law.

considered Jewish only if his mother is Jewish. Circumcision status has no bearing on affiliation with Judaism. Many Jewish men in many parts of the world are genitally intact and correctly consider themselves to be Jews. Consequently, protecting babies from circumcision in no way violates the right to religious freedom. Anyone who disagrees is obviously misinformed about Judaism and makes himself instantly suspect of ulterior, unconstitutional motivations.

7.2 The Fourth Amendment

The Fourth Amendment declares:

> The right of the people to be secure in their persons, houses, papers, and effects, against unreasonable searches and seizures, shall not be violated….

When a helpless infant is strapped to a circumcision restraining board, he is hardly "secure in his person against unreasonable seizures." The amputation and destruction of the penile foreskin is clearly a "seizure" of one's property.

7.3 The Fifth Amendment

The Fifth Amendment guarantees the right to due process and the right to life and property. It states, in part:

> …nor (shall any person) be deprived of life, liberty, or property, without due process of law; nor shall private property be taken for public use, without just compensation.

Circumcision clearly violates this amendment by endangering the child's life and depriving him of a part of his own body. Moreover, when hospitals harvest foreskins from newborn babies and then sell them to biotech, pharmaceutical, and cosmetic companies that use them as raw materials for products that generate millions of dollars of annual sales, it is the infant who deserves "just compensation." One may ask, however, precisely how much money is just compensation for depriving an American male of his full complement of normal penile function and sexual pleasure for a lifetime.

7.4 The Eighth Amendment

The Eighth Amendment in its entirety states:

> Excessive bail shall not be required, nor excessive fines imposed, nor cruel and unusual punishments inflicted.

While the founding fathers did not have circumcision in mind, since the medical profession did not adopt circumcision until the mid to late nineteenth century, they did want to protect citizens convicted of a crime from cruel and unusual punishments. To an innocent infant, circumcision, with or without anesthesia, is a cruel and unusual punishment without any justification.

7.5 The Ninth Amendment

If there is any question that cruel and unusual punishment applies to circumcision (Eighth Amendment), then that is clarified in the Ninth Amendment. The Ninth Amendment makes it abundantly clear that basic rights not specifically enumerated in the Constitution are still to be protected, and not denied and disparaged. The founding fathers established protections against circumcision more thoroughly than they could have imagined in the Ninth Amendment. Here it states:

> The enumeration in the Constitution of certain rights shall not be construed to deny or disparage others retained by the people.

Circumcision is perfectly described as "denial" and "disparagement." The victims of circumcision have been denied protection from bodily harm. Furthermore, they have been denied the ownership and use of part of their own body. Tangentially, employing the psychological use of the word "deny," we find the medical profession denying that it is perpetrating harm and, most disturbingly, the denial – born of tragic ignorance – by many circumcised males that circumcision has impaired them.

In order to maintain this denial in the face of overwhelming evidence to the contrary, disparagement is freely used. Many circumcised men are driven to engage in the ego-protecting cognitive dissonance of trivializing and disparaging the adverse effects of circumcision. They also disparage men who were fortunate enough to escape circumcision. Some circumcised males go to great lengths to avoid admitting to themselves the considerable harm that has been done to them. Denial and disparagement are hallmarks of circumcision.

8. RELIGIOUS CIRCUMCISION

Since autonomy and the right to keep all of one's body parts (one's property) is a basic human right, a privilege of United States citizenship, and constitutes immunity from seizure, then religious circumcision is illegal under the Constitution in the United States. Congress has yet to act on it. To date, the Supreme Court has not been presented with a test case.

Since circumcision is harmful, as the great Jewish rabbi and physician Maimonides made abundantly clear in the twelfth century, then Jewish infants, too, need to be protected as well as all other American infants.[6] An accusation of anti-Semitism might be leveled at this statement, but it is inappropriate for two reasons. Since scientific evidence proves that circumcision is harmful, Jewish boys also have a constitutional right to protection from circumcision. Jewish parents do not deny that they wish to protect their children from harm. Secondly, the proportion of activist Jews in the movement to eradicate circumcision is significantly higher than their proportion in the general population.

There is precedence for change within Judaism. Animal sacrifice, which was once an essential a part of Jewish ritual, is no longer practiced. It is important that individuals of the Jewish faith reconsider the practice of circumcision in the light of modern scientific information. Many enlightened Jews have already abolished ritual circumcision and have replaced it with a peaceful, non-bloody ceremonial *bris* without cutting (*Brit Shalom*).

9. THE POWER OF MYTH

To understand how doctors could torture and mutilate their infant patients through circumcision, one needs to understand the power of myth. Many doctors have actually convinced themselves that circumcision is painless, and that, at any rate, an infant does not remember it. They have adopted the self-serving view that the foreskin is "just a piece of skin" and that it is functionless — as if these conditions, even if they were true, somehow justified amputation and destruction. Nothing could be further from the truth. Circumcision produces dangerous levels of pain.[7, 8, 9, 10] It has now been demonstrated scientifically that circumcision produces higher pain responses in an infant than in an adult.[11] Studies also show that circumcised infants are more sensitive to pain months after the circumcision.[12, 13, 14, 15, 16, 17, 18]

The foreskin is the name given to the double-layered outer portion of the skin of the normal penis. It is an integral part of a complex organ. It has several primary functions: it protects the glans penis and urinary tract opening throughout life; it slides up and down the shaft like a piston to trigger pleasure and ejaculatory nerves; it unfolds during an erection to cover the elongated organ; and, via the ridged band, located on the inside of the foreskin, it contributes significantly to sexual pleasure.[19]

Far too many doctors labor under the false belief that circumcision is safe, that there are few complications, and that it has little or no adverse effect on later life. In fact, circumcision kills an undisclosed number of infants in America each year.[20, 21, 22, 23, 24, 25, 26, 27] In truth, the complication rate of circumcision is 100 percent because, in each instance, a normal, functioning body part has been removed for inadequate and bogus reasons. Many doctors believe that circumcision prevents cancer of the penis, but a preventive measure cannot ethically be employed unless the benefits outweigh the risks. In this case, the risk of getting penile cancer is so low that circumcision cannot possibly be justified as a preventive measure. Circumcision causes far more harm than anything it allegedly prevents.

10. THE VIOLATIONS OF DOCTORS

Doctors who circumcise may think they are doing the right thing, but only because they have failed to give the issue sufficient thought.[28] A doctor who circumcises violates the Golden Rule; he violates a primary tenet of medical practice, *First, Do No Harm*; and he violates all nine principles of the American Medical Association Code of Ethics.[29] He violates a basic human right of his patient – the right to an intact body. He violates the Nuremberg Code of Ethics, which was developed after World War II, as a direct result of Nazi experimentation on humans. The Code originally regulated experimentation on humans, but later evolved into a code of ethics for doctors. In view of these massive transgressions, a doctor who circumcises can no longer be considered a doctor.

Lawsuits have already been tried and settled on these issues.[30] The Constitution has been, and will continue to be, a strong ally for the plaintiff in these cases.[31] There is a compelling analogy with the scandal over the tobacco industry, which knew about the harm of cigarette smoking but publicly denied it for years. The industry's efforts at a cover-up failed and were soon followed by successful lawsuits. Insurers could easily save vast quantities of money by simply refusing to insure doctors who perform circumcisions. Doctors need to decide whether they would be better off abandoning circumcision or continuing to harm their patients, making money by fraud and, at the same time, risking their reputations by losing a lawsuit. It is up to the citizens and politicians of the United States, however, to decide how much longer they will allow the medical profession to continue violating human rights and the United States Constitution. The Constitution will eventually triumph over this misguided surgical abuse of innocent citizens, as it always has.

REFERENCES

[1] Denniston GC. Circumcision: An Iatrogenic Epidemic. In Denniston GC, Milos MF, editors. Sexual Mutilations, A Human Tragedy, New York: Plenum Press, 1997. pp. 103-109.

[2] Taylor JR, Lockwood AP, Taylor AJ. The prepuce: specialized mucosa of the penis and its loss to circumcision. Br J Urol. 1996 Feb;77(2):291-5.

[3] Cold CJ, Taylor JR. The prepuce. BJU Int. 1999 Jan;83 Suppl 1:34-44.

[4] Denniston GC. Circumcision and Sexual Pleasure. In Denniston GC, Hodges FM, Milos MF, editors. Flesh and Blood: Perspectives on the Problem of Circumcision in Contemporary Society. New York: Kluwer Academic/Plenum Publishers, 2004. pp. 45-53.

[5] O'Hara K, O'Hara J. The effect of male circumcision on the sexual enjoyment of the female partner. BJU Int. 1999 Jan;83 Suppl 1:79-84.

[6] Friedländer M, trans. The Guide of the Perplexed of Maimonides. 3 parts. New York: Hebrew Publishing Co.; 1881. part III. p. 267.

[7] Lander J, Brady-Fryer B, Metcalfe JB, Nazarali S, Muttitt S. Comparison of ring block, dorsal penile nerve block, and topical anesthesia for neonatal circumcision: a randomized controlled trial. JAMA 1997 Dec 24-31;278(24):2157-62.

[8] Sara CA, Lowry CJ. A complication of circumcision and dorsal nerve block of the penis. Anaesth Intensive Care 1985 Feb;13(1):79-82.

[9] Berens R, Pontus SP Jr. A complication associated with dorsal penile nerve block. Reg Anesth 1990 Nov-Dec;15(6):309-10.

[10] Snellman LW, Stang HJ. Prospective evaluation of complications of dorsal penile nerve block for neonatal circumcision. Pediatrics 1995 May;95(5):705-8.

[11] Anand KJ, Hickey PR. Pain and its effects in the human neonate and fetus. N Engl J Med 1987 Nov 19;317(21):1321-9.

[12] Taddio A, Goldbach M, Ipp M, Stevens B, Koren G. Effect of neonatal circumcision on pain responses during vaccination in boys. Lancet 1995 Feb 4;345(8945):291-2.

[13] Taddio A, Katz J, Ilersich AL, Koren G. Effect of neonatal circumcision on pain response during subsequent routine vaccination. Lancet 1997 Mar 1;349(9052):599-603.

[14] Taddio A, Stevens B, Craig K, Rastogi P, Ben-David S, Shennan A, Mulligan P, Koren G. Efficacy and safety of lidocaine-prilocaine cream for pain during circumcision. N Engl J Med 1997 Apr 24;336(17):1197-201

[15] Taddio A, Ohlsson A, Einarson TR, Stevens B, Koren G. A systematic review of lidocaine-prilocaine cream (EMLA) in the treatment of acute pain in neonates. Pediatrics 1998 Feb;101(2):E1.

[16] Taddio A, Ohlsson K, Ohlsson A. Lidocaine-prilocaine cream for analgesia during circumcision in newborn boys. Cochrane Database Syst Rev 2000;(2):CD000496.

[17] Taddio A, Pollock N, Gilbert-MacLeod C, Ohlsson K, Koren G. Combined analgesia and local anesthesia to minimize pain during circumcision. Arch Pediatr Adolesc Med 2000 Jun;154(6):620-3.

[18] Taddio A. Pain management for neonatal circumcision. Paediatr Drugs 2001;3(2):101-11.

[19] Taylor JR, Lockwood AP, Taylor AJ. The prepuce: specialized mucosa of the penis and its loss to circumcision. Br J Urol. 1996 Feb;77(2):291-5.

[20] Gellis SS. Circumcision. Am J Dis Child 1978 Dec;132(12):1168-9

[21] Baker RL. Newborn male circumcision: needless and dangerous. Sexual Medicine Today 1979 Nov;3(11):35-6.

[22] Racher P. Grand jury to probe death of baby after circumcision. Des Moines Register. (November 20, 1982):3A.

[23] Associated Press. Boy in coma most of his 6 years dies. The State (Columbia, South Carolina) (July 10, 1992):5B.

[24] Rogers P, Wallace R. Infant bleeds to death after being circumcised. Miami Herald (June 26, 1993):B1-B2.

[25] Lum L. SoRelle R. Boy's autopsy results expected: five-year-old lapsed into coma following circumcision. Houston Chronicle (July 28, 1995):28A.

[26] Associated Press. Circumcision that didn't heal kills boy: the boy's circumcision wasn't healing properly. News Net 5. (October 20, 1998): URL: http://www.newsnet5.com/news/stories/news-981020-051123.html.

[27] Foley JM. The unkindest cut of all. Fact 1966 Jul;3(4):2-9.

[28] Denniston GC. Tyranny of the Victims: An Analysis of Circumcision Advocacy. In Denniston GC, Hodges FM, Milos MF, editors. Male and Female Circumcision: Medical, Legal and Ethical Considerations in Pediatric Practice. New York: Kluwer Academic/Plenum Publishers, 1999. pp. 221-240.

[29] Denniston GC. Circumcision and the code of ethics. Humane Health Care International 1996;12(2):78-80.

[30] Llewellyn D. Penile Torts in the Courts. In Denniston GC, Hodges FM, Milos MF, editors. Flesh and Blood: Perspectives on the Problem of Circumcision in Contemporary Society. New York: Kluwer Academic/Plenum Publishers, 2004. pp. 69-79.

[31] Bonner C. The Activist's Rights in the Workplace. In Denniston GC, Hodges FM, Milos MF, editors. Flesh and Blood: Perspectives on the Problem of Circumcision in Contemporary Society. New York: Kluwer Academic/Plenum Publishers, 2004. pp. 81-88.

Chapter 18

TOWARD REGULATION OF NON-THERAPEUTIC GENITAL SURGERIES UPON MINORS

A Preliminary Legal Strategy

John V. Geisheker
Doctors Opposing Circumcision, Seattle, WA, USA

Abstract: An incremental, rational, inexpensive, risk-free plan is proposed for curbing circumcision. It avoids the risk of legal sanctions for filing frivolous lawsuits. Adapted for local use, this plan may hold the key to hobbling, and eventually eliminating, all unconsented genital mutilations.

1. INTRODUCTION

Non-therapeutic medicalized circumcision of infants and children in the United States, (where seventy-five percent of such surgeries worldwide occur), is now one-hundred-thirty-five years old, if we use the date medical historians use — 1870, the year of Dr. Lewis Sayre's notorious use of circumcision to "cure" a five-year boy of paralysis.[1] Clearly, the medical procedure itself existed prior to 1870, but likely it was mostly therapeutic.

The "medicalized" version of non-therapeutic circumcision, an Anglo-American invention, now affects an estimated 1.6 million boys a year worldwide, 1.2 million males in the United States, the major offender, and an unknown number of female victims.[2]

With such a head start, and without a single-payer medical system, it should come as no surprise that U.S. medical societies are resistant to change and will not voluntarily abandon the practice as summarily as did tiny New Zealand.[3] Not even, I suspect, to the lesser extent of other English-speaking countries — the United Kingdom, Canada, and Australia — where the

G.C. Denniston et al. (eds.), *Bodily Integrity and the Politics of Circumcision*, 203—217.

practice arose in the late nineteenth century but has seen steady decline in the last thirty years.[4]

In the United States, this industry now exceeds one-billion-dollars per year (not including costs for complications, repair surgeries, and resulting lawsuits, which adds an additional billion dollars), with deeply entrenched special interests, financial and cultural, even including a lively commercial market for infant tissue.[5] That creates a lot of momentum and predictable reluctance to evolve, even in the face of international disdain, growing domestic resistance, and the beginnings of legal attention.[6]

Unfortunately, the law is a lumbering, clanking machine and, even when engaged, a lagging indicator of public sentiment that, famously, rarely leaps to take the lead. However, the law is not without some potential to assist, as I assert here.

2. THE SHIFTING SANDS OF MEDICAL JUSTIFICATION

Unwittingly, medical societies have made a rod for their own backs by systematically stripping the practice of male circumcision of all its historical medical justifications, one at a time, leaving only the wavering, inconstant, laissez-faire court of familiarity. That is especially true in the United States, where medical societies have slowly switched gears, disingenuously and without shame, to claim that they are now honoring merely cultural urges.[7] (Urges, it has to be said, for a ritual they invented and marketed to parents using a litany of medical scare tactics).[8] This is a significant shift away from relying on gossamer-thin, but deftly presented, medical myths, and it has legal consequences.[9]

For one thing, abandoning medical justification leaves male circumcision legally unstable, hardly different from the cultural female circumcision so widely reviled and now forbidden.[10 11]

The only exception is the near continuous clamor of human papilloma virus (HPV) and HIV/AIDS protection, claims which uniformly ignore bioethical concerns and, thus, are vulnerable. (As one of my legal colleagues asks succinctly, “Is it ethically acceptable to carve up the genitals, at known risk, of one non-consenting person to hypothetically lower the risk of disease in an unidentified other — in the future — maybe?”)

The HIV campaign is little more than the modern equivalent of the tuberculosis protection claims of 1880, brought forward one-hundred-twenty-five years, exploiting exactly the same brand of desperation. It will be understood one hundred years hence, as vaccines are developed; indeed, that is already happening with (HPV).[12] Moreover, North America, with the developed world's highest percentage of circumcised males, about seventy percent, also has the developed world's highest rate of HIV/AIDS.[13] As a putative anti-AIDS public health panacea, circumcision, an unplanned mega-experiment, has already proved an ignominious failure

When, in 1971, the American Academy of Pediatrics declared the practice of male circumcision "unnecessary," and even including their suspicious backing and filling of the thirty-four years thereafter,[14] it was only a matter of time before some court would face the tough question: "Is a proxy consent for a non-therapeutic cultural surgery on a minor, sufficient to overcome the common law assumption that such occasions, lacking medical necessity, amount to battery upon the patient?"

At least one law professor who specializes in medical ethics has been bold enough to go beyond technical battery and state, forthrightly, that non-therapeutic circumcision is "criminal assault."[15]

Sadly, in the United States anyway, it may not yet be time to frame the question quite that provocatively for males, though it has already been answered for females, by federal law. But certainly it is time to inquire: What really *is* happening in the totally unregulated industry of infant (and vulnerable adult) genital surgeries, male and female, and the tawdry commercial trafficking in purloined tissue?

3. BRIGHT LIGHTS IN DARK CORNERS

About this dimly-lit corner of medicine, there is ample opportunity and existing channels for inquiry, official scrutiny, and perhaps, court-ordered investigation. Even a begrudging, half-hearted survey by medical authorities might serve to discourage the more egregious abuses.

The delivery of medical care in the United States is largely governed at the state level by two distinct, but cross-pollinated organizational tiers, a state board or department of health and (only nominally beneath) private medical societies.

State boards of health present themselves as vigilant protectors of the consuming public, with grand mission statements and impressive oversight machinery. In reality, they are staffed by or use as consultants the very same professionals they regulate, and they operate as much to shield assorted medical professions as to protect the public.[16] These state agencies may have statutory oversight duties, but they typically license and regulate professionals indulgently, and reprimand errant licensees reluctantly, after secret deliberations.

The private medical societies are worse — little more than medieval guilds, guarding the fiefdom of their members, erecting professional parapets, campaigning for legal immunity, and stabilizing earnings.

I freely confess, bar associations for lawyers are hardly much better; more's the pity. But to be fair, we lawyers, by profession, are less capable of doing serious damage to life, limb — or genitalia.

Moreover, there is substantial evidence at present of medical corruption and abuse, which, if well documented and artfully presented, might prove sufficient to chill the more outrageous practices of unregulated infant genital surgeries and begin a reform cascade. At a minimum, it might throw some much-needed public illumination toward what the courts have historically treated as a private decision between patient (or proxy) and the physician. At best, a lot of light might eventually lead to regulation.

And, of course, where public funds are expended, careful public oversight is reasonable and expected, and even legally enforceable by the courts.

Such scrutiny could never be more necessary than over the potentially abusive situation where the patient is a minor, the doctor has a financial stake in the outcome, and the proxy-consent parents are ignorant of the medicine. Or worse, the parents, neutral or puzzled to begin with, have been carefully groomed into co-conspiracy with the doctor, and, seduced by imagined cultural gains, disregard their child's best medical interests.

4. IDENTIFYING ABUSES AND THE CHALLENGE

Here is what I propose to initiate in the state of Washington, a challenge to my legal colleagues in other jurisdictions. My fond hope is that it may become a model for concerned individuals and organizations in other states, with lessons learned all around. It is clearly no silver bullet or wooden stake, and I make no claim other than that it is worth trying and poses, I argue here, minimal risk.

First we need to identify the existing abuses, which many bioethicists (but rarely the general public) already know by direct or anecdotal evidence:

- forced retractions or unnecessary lysing of the naturally adherent foreskin of intact infants, toddlers, and boys under eighteen, contrary to the best medical practice and specific proscription by pediatric societies,[17] but nevertheless, very common.

- circumcision of premature infants, against all reasonable precautions for their safety.

- circumcision of dying infants, where claims for future prophylaxis are patently dishonest.[18]

- scanty consents, both oral and written, which operate as marketing devices rather than duly diligent disclosure.

- fraudulent use of diagnostic code ICD-9-605, bogus infant "phimosis," an impossible diagnosis of pathology for neonates and young boys alike, but very common, found on millions of medical records.

- use of various CPT codes to circumvent medical assistance scheme restrictions.[19]

- use of bent, distorted, old, mismatched, or counterfeit clamps; and failure to inspect and sunset obsolete equipment, ignoring worldwide warnings by government agencies.[20]

- the 'iatrocycle' of botched infants, cases clinicians cover up and pass back and forth.

- harvesting and sale of purloined infant tissue for commercial use or stem-cell research, without disclosure or compensation.

- discouraging parents from pursuing the legal rights of the injured child as "traumatic for your boy" (until the statute of limitations runs).

- nursing home circumcisions of elderly or paralyzed males no one cares to bathe properly.

- circumcisions of adults for invented reasons — phimosis, paraphimosis, and frenulum breve — without more conservative care advised or attempted.

It is better to cite specific examples of such occurrences, with individual patient and physician names and details and supporting medical data, but, failing that, records showing public funds expended on fraudulent procedures and those to repair botches, for instance, should themselves be enough to trigger an inquiry.

United States federal law, the Health Insurance Portability and Accountability Act of 1996 (HIPAA), a stringent codification of the rights of the patient to medical record privacy, may hinder easy record access and provide some cover for institutional fraud. A flanking maneuver, though, using evidence of public funds squandered, should obviate the need for a frontal assault. There is even the possibility HIPAA could be of some use.[21]

5. COMMON FRAUDS AND DODGES

Public laws allowing public document access and disclosure can be used to unearth all sorts of financial abuses.[22]

A classic example of rampant fraud is one that is widespread, especially among those fifteen US states that have defunded circumcision. In fiscal 2002, practitioners in Washington state, where circumcision has been defunded for many years, used diagnostic code ICD-9-605 ("congenital phimosis and redundant foreskin") on 902 "occasions" (i.e., 902 real, actual children), to request over $400,000 from state and United States federal Medicaid funds. Similar amounts have been claimed in prior years.[23]

This is pure, inexcusable fraud, a flagrant violation of federal and state law, and a transparent attempt to extract a fee, the best interests of the child notwithstanding. (A much larger sum would have been expended by private insurers and HMOs, paying reflexively under ICD-9-605 and dubious CPT codes, for care provided boys whose parents were defrauded into thinking their child required surgical repair for normal, developmental anatomy. While obvious fraud, this is a private matter, not a public coffer concern.)

6. PUBLIC FUNDS EXPENDED FOR MINORS WITH IATROGENIC INJURIES

In Washington state, some thirty-five percent of boys are circumcised each year, over 14,000, a relatively low rate in the United States and dropping, even as levels in more populous Midwest states are stable at seventy percent and above. Statistically, some of those Washington boys will need a second or even third procedure paid for by the public. Using federally subsidized funds to repair iatrogenic damage, appears, ironically, entirely legitimate, though these costs should more properly be the responsibility of the defendant doctor's medical malpractice carrier.

Reliable projections suggest at least ten percent of boys will suffer complications of some sort, as studies show follow-up care can range as high as fifty-five percent.[24] (Doctors Opposing Circumcision takes the position that circumcision has a hundred percent complication rate — risk and damage incurred without therapeutic justification. But of course, the popular misconception is that the procedure is benign, if unnecessary and merely elective.)

Indeed, a recent study showed twenty-nine percent of boys were seen for iatrogenic, post-circumcision adhesions, though some of these would likely resolve spontaneously.[25] Thus, using even a conservative ten percent complication rate, 1,400 Washington boys will require extra follow-up care, a boon practitioners are likely loathe to forego, and a drain, where Medicaid funds are tapped, on ever more strained public resources.

Of that 1,400, some one hundred or so boys each year will be seriously injured, enough to require hospitalization and surgical repair under general anesthesia at a cost of US$5 to $15,000 per patient. One leading pediatric urologist in the Seattle area alone proudly claims to do fifty such circumcision repairs, partial phalloplasty, meatotomy, or skin graft surgeries each year. Other urologic surgeons might do fewer, but collectively, at least, one-hundred boys are seriously injured in the Puget Sound region annually. It is likely that careful scrutiny of other United States urban areas, with equivalent populations, would produce similar or higher numbers.

7. LEGAL REDRESS DISCOURAGED

Predictably, few within medicine are inclined to expose this 'iatrocyclic' and lucrative scenario. It is also unlikely that urologists who do the botch repairs are reporting, to the appropriate medical oversight authorities, those obstetricians, pediatricians, and family practice doctors who do substandard work or aggressively market unnecessary surgeries. What possible incentive would they have to anger their physician colleagues upon whom they rely for referrals?

Because of the cost, nuisance, and what was likely a whimsical choice anyway, the individual family is likely to feel embarrassed — and without advice to the contrary — uniquely unlucky. Private conversations this author has had with families and medical professionals involved indicate physicians discourage parents from pursuing any legal remedy, alleging it would be traumatic and embarrassing for the family. In any case, why bother when the only likely legal claim, assuming the family had medical insurance, is one for pain and suffering to a toddler, with his future voiding and sexual difficulties merely speculative and a long time hence?

As one might expect, the small community of urologists uniformly refuses to testify against their obstetric and pediatric colleagues in such cases. One urologist whom I know personally, who claims to discourage infant circumcision and believes parents who opt for it are foolish, shrugs and says, "Well, the parents took their chances." Likely, in the present

- Reporting by nursing homes on circumcisions of the vulnerable, elderly, or paralyzed. (These are usually to accommodate staff who did not or will not properly bathe a resident.)

- Tracking cases of forced gender reassignment and female circumcision to monitor for cultural surgeries masked as therapeutic medicine.

Of this last, I believe state, federal, and even extra-territorial laws forbidding female circumcision could be of some use to concerned male circumcision opponents and forced gender reassignment opponents. Female circumcision victim-advocates have been notably reluctant to link their concerns with that of male circumcision, which they characterize as comparatively benign, apparently for strategic reasons. But that should not discourage male circumcision opponents from suggesting that female circumcision may be continuing *sub rosa* and thus require scrutiny.

The so called 'Seattle Experiment' of 1996 provides clues as to how female circumcision violations might be occurring despite United States federal law:

In 1996, physicians at a large public hospital in Seattle offered to perform modified *sunna* circumcisions, a clitoral hood incision, using sterile technique, on Muslim girls.[27] Their argument was that this would save the child from the infection risk and the family from the burden of a round-trip to Africa or the Middle East, for a more radical, septic version by tribal circumcisers, the so-called "summer vacation circumcision." The plan was loudly criticized as an attempt to medicalize and normalize an unfamiliar, non-therapeutic, cultural practice, and thereby endorse and sustain female circumcision. (Little was mentioned about ongoing, non-therapeutic, also cultural, male circumcision.) Around the same time, United States federal law, forbidding all such non-therapeutic, cultural procedures on females rendered the Seattle experiment moot.[28]

It would be a simple matter, however, for an errant United States doctor to accommodate the immigrant parents by claiming the child had atypical genitalia and that the surgery was thus 'therapeutic,' evading the reach of the law. Histopathologic evidence to the contrary would be, as usual, spoliated or destroyed.

'Therapeutic,' of course, is the same shield used by those who rush forced gender reassignment on a child with ambiguous or exceptional genitalia, using arbitrary esthetic standards rather than documented neurophysiological, endocrinological, and psychological assessment.

That temptation toward forced gender reassignment or female circumcision by itself provides some support for concerns that, at a minimum, the paying public has a right to know what is happening. How many infant genital surgeries are being performed, on whom, for what justification, and at what cost in public treasure? This is true, whether for initial surgeries, male or female, or the cost of repairing unnecessary interventions gone awry.

9. POSSIBLE REFORMS TO URGE

As a second salvo, the complaint might take the form of suggesting possible reforms that would bring Washington state medicine more in line with other world medical-society standards for similar procedures. Substantial reforms, for instance, have been urged by the governing College of Physicians and Surgeons in neighboring British Columbia, Canada, following pressure from the public and the death, by hemorrhage and hypovolemic shock, in 2002, of a one-month old boy, Ryleigh R. B. McWillis.[29] Such reforms might include:

- A uniform proxy consent (that bioethicists could write or edit).
- Mandatory reading and signing of the state statute on proxy consent (requiring, in Washington, recitation of the standard of "likely preference and best medical interests of the patient") by both doctor and parent.[30]
- Parents to view a film of the common, no-anesthesia, Gomco circumcision.
- Signature of both parents to prevent both "spite circumcisions" and the presentation of proxy consent forms to the mother while she is under medication or stress.
- Mandatory histopathology for all genital surgeries to boys or girls under eighteen years of age.
- Photographic and genetic evidence with second opinions and hospital bioethical review for all forced gender reassignment surgeries.
- A four-hour professional observation period for every outpatient circumcision, to detect bleeding and prevent hypovolemic shock for infants in the care of non-medical personnel — like parents. This caution recognizes it can take as little as 2.5 ounces of blood lost to kill a large infant.[31]
- Graphic demonstration of how small a stain the lethal 2.5 ounces of blood creates in a modern disposable diaper treated with absorbing agents.
- ICD-9-605 and other code fraud exposed and flagged, where public money or insurance is involved.

- A recitation of the legal exposure of both physician and parent when the child reaches the age of majority.
- Tissue sales to require a separate proxy consent and payment into a trust for the child.
- Mandatory Continuing Medical Education (CME) credits on care of the intact (non-circumcised) child.

10. THE ANESTHESIA IRONY

Careful readers will note that I have not included the serious bioethical and psychological issue of withheld anesthesia. Although the better clinical practice is to use injected local anesthesia, and safe protocols have existed and been urged for decades, its use is still regrettably rare for circumcisions. Ironically, guidelines for lab animal surgery and veterinary practice require effective anesthesia (in the United States, by law), and forbid mechanical restraints in lieu.[32] Infants have only the protection of ethical medical practitioners, not much in evidence with a circumcision performed behind closed doors upon a wide-awake infant strapped to a plastic tray. Moreover, even injected anesthesia is not completely effective for males, as their complex genital innervation resists both topical (EMLA) and shallow local anesthesia, and general anesthesia for infants holds well-recognized risks. The *legal* risk, to quote a biblical legal aphorism, is that the medical authority or a court might "cut the baby in half," awarding anesthesia to children as a humane, generous compromise. This apparently sweeping "advance" might sanitize genital surgeries, overshadowing the underlying, more fundamental ethical concerns of whether they should have been attempted *ab initio*, in the first instance. Experienced attorneys know that one does not structure one's case to tempt a judge to an easy, popular, solution, leaving the fundamental issue side-stepped and unstable. Hence my reluctance to press for anesthesia too early unless there is already headway on other issues.

11. CONCLUSION AND RISKS

Finally, the reader will surely have seen our reasoning, similar to that used by anti-abortion activists: banning cultural, non-therapeutic, infant genital surgeries outright is politically impossible, but making them logistically more cumbersome, time-consuming, less lucrative, even embarrassing, might prove less controversial, and eventually, more effective.

More to the point, a court is likely to agree that the enumerated financial abuses cry out for some scrutiny and even reform, where they would never ban circumcision outright based on mere bioethical arguments. Without the financial abuses exposed, United States local courts would fear public outcry by a coalition of medical practitioners losing trade, immigrants with a cultural stake in the outcome, and those alleging infringement of religious freedom.

I freely concede, aside from being optimistic, this plan is vulnerable to criticism that it might, like the "Seattle Experiment," merely medicalize the existing situation and thereby provide legal sanction and cover. I believe such a risk is worth the gamble. I do not think any particular occasion of male infant circumcision is lucrative enough to sustain the burden I propose. I believe *any* scrutiny may be enough to discourage those many practitioners not deeply invested in the practice anyway — and send a warning to others. Indeed, there are those sympathetic medical professionals who claim to need an excuse to discourage parental cultural whims and who might welcome "cover" provided from without. Those circumcisers with a deep-seated psychological need to do so, could not, alas, be curbed even by an outright legal ban. Fortunately, they are likely a minority.

Finally, I believe this plan, tailored to each individual jurisdiction, is a risk-free first step. Adapted for local use, this plan may hold the key toward hobbling, and eventually eliminating, all non-therapeutic, merely cultural, nonconsented genital mutilations of minors, male or female, and the financial abuses and chicanery that seem, inevitably to attend them.

REFERENCES

[1] Gollaher DL. From ritual to science: the medical transformation Of circumcision. America Journal of Social History 1994;28(1):5-36.

[2] Bollinger D. Normal versus Circumcised: U.S. Neonatal Male Genital Ratio, 2004, white paper, February 12, 2004. An original publication by the Circumcision Information Resource Pages. February 2004. http://www.cirp.org/library/statistics/USA/.

[3] McGrath K, Young H. A review of Circumcision in New Zealand. in: Denniston GC, Hodges FM, Milo MF (eds). Understanding Circumcision: A Multi-Discliplinary Approach to a Multi-Dimensional Problem. New York: Kluwer Academic/Plenum Publishers, 2001.

[4] http://www.cirp.org/library/statistics/

[5] Fleiss PM. The case against circumcision. Mothering (Winter 1997):36-45.

[6] A good discussion of the increasingly pointed legal scholarship and rising bioethical concerns may be found within: Waldeck SE. Using male circumcision to understand social norms as multipliers, University of Cincinnati Law Review 2003;72(3):455-526.

[7] Cf., for instance, American Academy of Pediatrics: Task Force on Circumcision. Circumcision policy statement. Pediatrics 1999 Mar;103(3):686-93.

[8] Gollaher DA. Circumcision: A History of the World's Most Controversial Surgery. New York: Basic Books, 2000. p. 92.

[9] Giannetti MR. Circumcision and the American Academy of Pediatrics: should scientific misconduct result in trade association liability? Iowa Law Review 2000 May;85(4):1507-68.

[10] 18 United States Code Part 1, Chap 7, §116; similar laws have been passed in most other First-World countries.

[11] A covenant with the status quo? Male circumcision and the new BMA guidance to doctors, J.Med. Ethics 2005;31;463-469.

[12] http://www.rrpwebsite.org/hpvvac1.htm.

[13] Centers for Disease Control and Prevention: HIV/AIDS Surveillance Report 2001;13(No. 2), HIV/AIDS Surveillance Report 2002;14 and HIV/AIDS Surveillance Report 2003;15.

[14] http://www.cirp.org/library/statements/

[15] Somerville MA. The Ethical Canary Science, Society and the Human Spirit. Toronto: Viking/Penguin, 2000. pp. 202-19.

[16] Washington State's Medical Quality Assurance Commission (WMQAC) is typical, with 13 physicians, 2 physician assistants, and only 4 members of the public aboard. See: https://fortress.wa.gov/doh/hpqa1/HPS5/Medical/commission.htm.

[17] "Forcing the foreskin to retract before it is ready may severely harm the penis and cause pain, bleeding and tears in the skin." Newborns: care of the uncircumcised penis: Guidelines for parents. Springfield: American Academy of Pediatrics, 1990.

[18] Doctors Opposing Circumcision is appealing a formal rejection of one such complaint against the hospital, physician and those assisting: "The death of Justin William Rumberger, DOD 12/28/03; Washington State Department of Health, (WSDOH) case: 2004-04-0014MD."

[19] Examples include: CPT54150, CPT54160, CPT54163. Unfortunately, mere existence of the codes appears to suggest possibilities of imagined pathology and intervention opportunity. Anesthesia codes 00920 and 64450 are also used to evade Medicaid restrictions on the underlying intervention, whether anesthesia was actually provided or not.

[20] FDA MedWatch Reports, July 1992 through January 2000. IPM Procedure: Circumcision Clamps, Health Devices 2000 January;29(1):22-3Hazard: Damaged Allied Healthcare Products Gomco circumcision clamp. Health Devices 1993 Mar;22 (3):154-5. Hazard: Incompatibility of different brands of Gomco-Type circumcision clamps. Health Devices 1997 Feb;26(2):76-7.

[21] In general, HIPAA preempts state law provisions that are contrary to its provisions, but allows exceptions: "(a) to prevent fraud and abuse related to the provision or payment for health care, (b) to ensure appropriate state regulation of insurance and health plans." Thus HIPAA has some potential to circumvent more restrictive state laws.

[22] These laws are variously termed public document access, public disclosure, sunshine laws and freedom of information act (FOIA), requests.

[23] The author received this information pursuant to a demand for disclosure under Washington law, RCW 42.17.260, the Public Documents Access law, after repeat and insistent written requests.

[24] Van Howe RS. A Cost-Utility Analysis of Neonatal Circumcision. Med Decis Making. 2004 Nov-Dec;24(6):584-601.

[25] Van Howe RS. Variability in penile appearance and penile findings: a prospective study. Br J Urol. 1997 Nov;80(5):776-82.

[26] RCW 26.44.030(1)(a) When any practitioner, ... has reasonable cause to believe that a child has suffered abuse or neglect, he or she shall report such incident, or cause a report to be made, to the proper law enforcement agency or to the department as provided in RCW 26.44.040.

[27] Ostrom CM. Harborview debates issue of circumcision of Muslim girls. Seattle Times (13 Sept 1996): P. A15.

[28] 18 United States Code Part 1, Chap 7, § 116.

[29] Newell TEC. Judgement of inquiry into the death of McWillis, Ryleigh Roman Bryan. Burnaby, B.C.: British Columbia Coroner's Service, Monday, 19 January 2004.

[30] RCW 7.70.065, Informed consent – Persons authorized to provide for patients who are not competent ...(3) Before any person authorized to provide informed consent on behalf of a patient not competent to consent exercises that authority, the person must first determine in good faith that that patient, if competent, would consent to the proposed health care. If such a determination cannot be made, the decision to consent to the proposed health care may be made only after determining that the proposed health care is in the patient's best interests.

[31] Nathan DG, Oski FA. Hematology of Infancy and Childhood. Philadelphia: WB Saunders, 1987. p. 1687. Smart J, Nolan T (eds). Paediatric Handbook. 6th ed. Carlton South, Victoria, Australia: Blackwell Science, 2000. p. 2.

[32] 7 United States Code 54, § 2143.

Chapter 19

STRATEGIES FOR LITIGATION

David J. Llewellyn
Atlanta, Georgia

The conundrum of a continuing high male neonatal circumcision rate in the United States in the face of the official opposition, or at least neutrality, of the American Academy of Pediatrics since 1970,[1] and despite increasingly vocal protest by medical,[2] psychological,[3] legal,[4] and lay organizations,[5] is puzzling to many at home as well as to most abroad. Americans' insistence on circumcising their newborn sons must seem particularly strange to those who reside in Western Europe, most notably in Italy and France where, since the Renaissance, there has existed a tradition of openly portraying the intact male nude in all artistic media, including monumental sculpture that graces public squares,[6] and where the museums are filled with ancient art depicting the intact male. Even well-written books outlining the specious reasons for the adoption of routine circumcision and the mental and physical disadvantages of its infliction on infants have been unable to induce a majority of American parents to forego circumcising their baby boys.[7, 8, 9] In short, circumcision is now a cultural tradition deeply embedded in the American psyche. It has become a cultural and social norm. Its status as such affects, to a great extent, its treatment by the American legal system.

Circumcision has been the predominant social norm in the United States since at least 1940. For the upper classes, it has been such since at least 1910. Encouraged by well-known nineteenth-century physicians, such as orthopedic surgeon Lewis Sayre in New York and Italian-born (Turin), but Minnesota-raised, Peter C. Remondino,[10] a graduate of the prestigious Jefferson Medical College of Philadelphia, circumcision was first touted as a remedy for masturbation, which was thought to have deleterious effects upon the male body.

Slowly, but surely, other ailments were added to the list of those supposedly prevented by circumcision until the American medical

G.C. Denniston et al. (eds.), Bodily Integrity and the Politics of Circumcision, 219—232.

establishment and nearly all parents were convinced of circumcision's absolute necessity. By the time of my birth in 1950, circumcision was so well entrenched in American medical practice that only premature infants, children of recent immigrants, and those whose parents were too poor to afford the cost of the operation avoided the knife at birth. So pervasive was circumcision that gentile boys generally believed they were born without a foreskin and that the possessor of one had a birth defect. I recall that, when I was in the United States Navy in the early 1970s, a fellow officer, a graduate of the United States Naval Academy, asked me why he could not see a difference between himself and his Jewish friends in the shower. Of course, the reason was that he was circumcised and so were they. He just didn't know what circumcision is. He apparently had never seen a Greek or Roman statue or had assumed that the human penis had changed since ancient times. Such ignorance is quite general in a society where public male nudity is virtually unheard of, even in art. A recent article on circumcision in a national parent's magazine had the author admitting that, since he was a Roman Catholic, he assumed he was not circumcised until his mother told him differently, when his wife was expecting their first child. He commented, "In the midst of deciding whether to circumcise Jordan, a somewhat embarrassing call to my mother set the record straight. Not only was I circumcised, but every boy I knew as a kid was, too."[11] This common ignorance was and is aided by sex instruction books for children that almost always show a circumcised penis or at least one that appears circumcised. In the 1950s, one well-known and widely read book on human anatomy for children completely ignored the subject by showing only a sagittal view of the male genitalia and by ending the penis just before the glans, thus avoiding having to explain that most boys had been surgically altered at birth[12] (Figure 1). Small wonder that my fellow officer did not know that he had been mutilated.

The situation was not much better in medical schools, nor is it much better today. As an exhaustive study by Gary Harryman has shown, many, if not most, medical texts in the United States do not correctly depict the normal penis.[13] One text that illustrates the Tanner stages of human maturation depicts the penis so that it appears that a boy grows a foreskin as he matures.[14] Physicians seem universally ignorant of normal foreskin

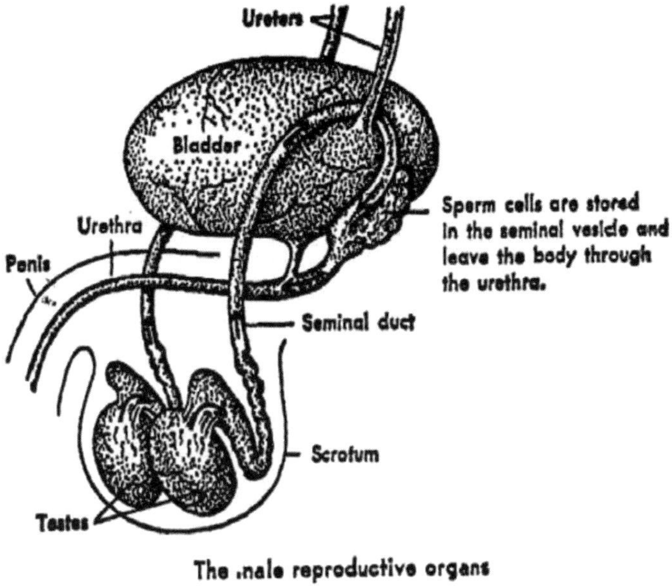

Figure 19-1. Inaccurate anatomical illustration in: Glemser B. All about the human body. New York, NY: Random House, Inc.; 1958. p. 131. Reproduced with permission.

development. Twenty-two years ago, my wife and I left our son's first pediatrician because he insisted that his foreskin should retract at three months or be forcibly retracted.[15] At the time, there was ample evidence that this was untrue. Today there is even more.[16] Nevertheless, I am called, on a monthly basis, by a distraught mother whose young son has had his foreskin prematurely and forcibly retracted by an ignorant American physician or nurse. Indeed, the ignorance seems almost willful, as though the doctors need to find a pathological condition in the normal foreskin in order to justify the social norm of removing it at birth.

The strength of the social norm, even in the face of correct information, can be appreciated from the following incidents. Several years ago, I represented a boy who had suffered a glanular injury when he was circumcised at birth. Despite this tragedy, the parents had his newborn brother circumcised.

Recently, an acquaintance of mine was pregnant with her first child. I gave her husband a great deal of information about the disadvantages and risks of circumcision. They were both aware that my sons are intact. They know that each of them was popular and successful in school. Nevertheless, when the mother first showed me her newborn son she immediately changed his diaper in front of me, apparently so I would see that she had rejected my advice and had him circumcised. She mentioned to me that she had been under a great deal of "pressure," but did not indicate if it was from relatives or from her own sense of social conformity.

Finally, several years ago I represented a young man who sued the physician who had circumcised him at his birth, 18 years earlier.[17] The essence of the suit was that my client's mother did not want to have him circumcised but had been induced to sign a consent form for the procedure while under the influence of drugs for the Cesarean section that she had just undergone. Thus, her consent was invalid and my client's circumcision, in law, was considered to be the commission of the tort, or civil wrong, of battery. In deposing the defendant doctor, I discovered that he had been born in New York of Italian immigrant parents, had returned to his family's farm outside Bari at an early age, and had earned his medical degree in Rome. He had returned to the United States as a young man and had done his residency in obstetrics and gynecology at a New York Roman Catholic hospital, where he apparently learned the American art of circumcision. He denied knowing that most Italian men are intact. He would not admit that the foreskin had any function whatsoever or that it greatly added to sexual pleasure. He admitted that he never gave his mothers any information in regard to circumcision unless they asked, but merely made inquiry as to their wish in that regard and then carried it out. When asked why he thought he had a moral right to circumcise an infant, he could not answer. From the look in his eyes, it was apparent to me that he had never thought about the issue before, but had merely followed the American social norm. Only now was he faced with the horrors he had committed countless times in the name of social conformity; and those horrors seemed to overwhelm him, an otherwise compassionate man and physician.

These examples demonstrate better than any argument the strength of social norms upon the circumcision rate in the United States. Contrary to popular opinion and Hollywood movies, the United States is not made up of social revolutionaries or independent thinkers. Indeed, one of the hallmarks of our culture seems to be social conformity. Those who stray from the well-trodden path in our democratic society risk social opprobrium. There is overwhelming pressure to conform to what most think of as "American" behavior. We want our sons to be fully "American." We want them to be tough, as well as conforming. Thus, our schools emphasize football over academics, and our landscape is pockmarked with chain restaurants, which are the same from coast to coast. Our purchasing habits are regulated by advertisements. Our citizens, whose grandfathers came from Greece, Hungary, Poland, Italy, and other countries, are entirely ignorant of their ancestral languages and cultures. Our boys are circumcised despite our knowledge of its pain and harmfulness. Our culture insists on conformity and exacts a price from those who challenge it. Small wonder that circumcision, which not only affects the sexual performance of a man but, by its very nature, defines his sense of the normal penis, is hard to eradicate. Likewise it should come as no surprise that law follows culture and that legal challenges to circumcision have heretofore been met with suspicion if not derision.

Recently, legal commentators have begun to consider the effects of social norms on the US legal system. An entire body of legal literature has emerged from the integrated study of law and medicine, law and psychology, and law and economics. Social norm theory is hotly debated in the academic legal literature. Two such recent articles discuss at length the problem of circumcision as an American social norm. Both posit that the American legal system will not begin to treat circumcision as it should be treated until such time as the social norm has shifted to non-circumcision. This has important implications for the success of legal challenges to circumcision, as my recent experience in a variety of cases proves.

In "Circumcision: Cultural-Legal Analysis"[18] recently published in the *Virginia Journal of Social Policy and the Law*, Professor Geoffrey P. Miller of New York University looks at the three periods of circumcision in American society. The first, when most American boys were intact, was prior to the late nineteenth century. The second was from the end of the nineteenth century to the end of the twentieth, during which time he contends "physicians acting as norm entrepreneurs reconceived the phallus. They portrayed the uncircumcised penis as polluted, unnatural, harmful, alien, effeminized and disfigured, and depicted circumcision as true, orderly, and good." He analyzes cultural attitudes on the bases of purity and pollution, health and harm, self and other, natural and unnatural, beauty and deformity, male and female, order and chaos, good and bad, and true and false. He posits that, while those espousing circumcision have been able to culturally craft a view that circumcision enhances purity, health, and self-image, and that it is beautiful, male defining, orderly, and good, and that the reasons given for it are true, the anti-circumcision movement of the last twenty or so years is tipping the scale in the other direction in all areas. He predicts that "medical circumcision of boys, even in the United States, will soon no longer be normative" and that "circumcision as a medical procedure . . . appears to be headed the way of tonsillectomies and other once popular but now unfashionable surgical interventions." However, he finds it "much more problematic...whether these [anti-circumcision] advocates will be able to embody their beliefs in legal prohibitions that parallel the existing legal norms against female genital mutilation." While he predicts "Unauthorized circumcisions are unlikely to escape legal sanction in the future, as they have in the past, on the ground that the operation was performed without malpractice," that "courts are likely to insist on informed consent practices that equal those commonly used for other surgical procedures," that "when damages are awarded, they are likely to be more substantial than awards in the past," and that anesthesia may come to be required by the standard of care, he does not foresee either the courts or the legislature outlawing neonatal circumcision itself. In this regard, he may well be correct as a practical matter because, as suggested in dicta by one Federal Court of Appeals judge, a "no-circumcision law would abridge both the right to rear one's children in one's religion [with reference to Jews and Muslims] and the free exercise of one's religion," and would be found to be in violation of the United States Constitution.[19]

On a similar note, Associate Professor Sarah E. Waldeck of Seton Hall University School of Law argues that social norms act as what she terms "multipliers," influencing "a person's perception of each and every factor in the behavioral calculus, including the costs and benefits of norm compliance." She opines, "The multiplier effect of norms is similar to a number of phenomena that cognitive scientists collectively refer to as 'confirmation bias,' or our tendency to seek information and ask questions that will corroborate rather than falsify our theories, and to interpret evidence in a way that supports our beliefs or hypotheses."[20] This certainly helps to explain why circumcision continues to be supported by doctors to reduce, by insignificant amounts, the incidence of certain diseases notwithstanding the very real danger of infection and physical disfigurement by way of accident that any such procedure entails. She posits that, so long as circumcision is seen as conferring the distinction upon parents that by circumcising their son they have been "good parents," the law will not greatly interfere with the practice. She rejects the notion that sudden change will occur just because anti-circumcisionists have the upper hand medically and morally. She suggests that a gradual approach, such as passing state laws prohibiting health insurance companies from covering routine neonatal circumcision, requiring pain control when circumcision is performed, and requiring true informed consent counseling before the procedure as a way to shift the cultural norm away from circumcision equals good parenting, to non-circumcision equals good parenting.

Miller's and Waldeck's observations ring true in my recent experience in disputes between parents, one of whom wishes to circumcise and the other of whom does not. Most often, this dispute arises between divorced parents where the parent with custody wishes to circumcise an older child who was not circumcised at birth due to the opposition of the parent who now does not have custody. While the obvious solution is to require both parents' assent to such a painful, invasive, and harmful procedure, virtually all American states give the parent with physical custody the sole right to decide "medical" issues or at least the right to decide in the case of a dispute. Most, if not all, American states allow a physician to proceed to operate on a child with the consent of only one parent when the parents are married. This often leads to "spite" circumcisions and real tragedy for the child.

For example, earlier this year, I was consulted by a young man who was separated from his Muslim wife, a native of Morroco. She had sole custody pending the divorce, but joint legal custody had been agreed upon to take effect at the time of the divorce. Knowing of his strong opposition to the circumcision of his son, the wife had the boy circumcised a few days before the divorce became final and failed to tell the physician of the dispute. Apparently the doctor had no problem circumcising a healthy three-year-old just because the mother wanted it done.

In that case, of course, the courts were not involved. However, in Baltzley v. Baltzley, a Pennsylvania case,[21] the Court sided with the custodial father against the non-custodial mother in a dispute over the threatened circumcision of their five- and eight-year-old sons. Despite being presented with undisputed testimony of some five hours in length, detailing the function of the foreskin, the damage caused by circumcision, the psychological effect of circumcision on boys of their ages, the risks of general anesthesia, the risks of surgical accident and infection, and the harm to the boys' future sexual lives, the Court refused to grant an injunction against the father prohibiting the boys' circumcisions. One of the reasons the father wished to have the boys circumcised was "to control urinating problems!" The Court held that the father "as the parent having sole legal custody of the two subject minor male children, has the legal right to make medical decisions in the best interest of the subject children after a medical examination of the subject children by qualified physician(s) and consultation with them of the benefits, if any, of the possible circumcision and the risks involved with the same." In other words, the Court refused to interfere in what it viewed as a private medical matter between the father and the physician. The case is on appeal.

Likewise in Azar v. Azar,[22] a Missouri case, the Court declined to enter an emergency order prohibiting the custodial father from circumcising the two-and-three-quarter-years old boy, despite the objection of the noncustodial mother, who had had custody of the boy during the first twenty-nine months of his life. The guardian ad litem appointed for the child opposed the emergency application on the ground that the father was given sole legal and physical custody and had been awarded the sole right to provide proper health and medical care for the boy.[23] The mother appealed the ruling, first to the Court of Appeals, which affirmed, and then to the Supreme Court, where the matter is still pending. The mother also wrote a letter to the risk managers at all local hospitals and to all urological groups in the United States indicating her opposition to the circumcision of her son and threatening suit if a circumcision should be performed. While the matter was pending on appeal the father found someone, presumably a physician, to circumcise their son. Thus, in this case, as in Baltzley, the Courts rested on legal technicalities rather than on the best interests of the child. In so doing, they conformed to the social norms that equate good parenting with circumcision and bad parenting with non-circumcision, thus confirming the theories of Miller and Waldeck.

The Azar case had another disturbing factor, which confirms how deeply entrenched pro-circumcision bias is in the United States. Ms. Azar received several obnoxious and abusive letters and telephone calls, either from physicians or someone purporting to be one. Illustrative of these communications is one in which a urologist, Frank R. Bacque, MD, of Lafayette, Louisiana, stated: "I am in receipt of this bizarre letter of yours, threatening to sue any urologist that would do a circumcision on your son, E., who is now approximately three years of age. I have been doing urology for the past 30 years, and I can assure you that the complications and problems of phimosis in later life are much greater than the 'irreparable harm' that some people, such as yourself, cite as ample medical evidence. This is all absolute foolishness." He concluded his harangue by stating that a prominent local family had her family name and that "[t]his sort of thing, under the guise of caring about your son, is dragging your wonderful name through the gutter."[24] Another urologist from Texas wrote, "Thank you for your recent thinly disguised rantings concerning your anticircumcision [sic] agenda. I must assure you that you and your family are not welcome at any time in my practice. I also will assure you that I am not responsible for any falsification concerning the identities of you, your son, or your ex-partner. Please be advised that I do not intend to spend any time 'on the lookout' for you or your family. I hope this concisely relates my thoughts on this matter."[25]

While not all cases end as badly as those just cited, the decisions are illustrative of the difficulty any parent has in bucking the American cultural norm of neonatal circumcision. All that a litigant can do is to present a compelling factual case through a knowledgeable and personable urologist and hope for the best. While such often succeeds, it sometimes does not, as these examples show.

Another example of the refusal of the courts to go out on a limb is the case of Haynes v. Smith.[26] In that Georgia case, a young man sued his circumciser for damaging him during his neonatal circumcision. The doctor had used a non-standard method to perform the circumcision, which involved using an electrocautery device to sever the foreskin rather than a scalpel. In the process, he removed a portion of the ventral glans. He did not tell the parents of his error. Haynes grew up believing that he had a normal penis. He discovered differently through research on the Internet during college. He came to me and we filed suit for medical malpractice. We also alleged battery on the ground that his mother had not had circumcision explained to her even in general terms before she gave consent and that she did not otherwise know what circumcision was as she had never seen an intact male. Finally, we alleged that, as circumcision is non-therapeutic and does not meet the definitions of medical or surgical treatment under Georgia law, his mother's consent was invalid and the procedure was a battery rather than malpractice.

The Georgia Court of Appeals applied the ten-year statute of repose, which is applicable only to medical malpractice actions, and affirmed the trial court's grant of summary judgment.[27] While Haynes had pleaded that the silence of Smith, after he had harmed him, amounted to fraud, which acted as an estoppel (bar) to the imposition of the statute of repose, the Court accepted Smith's testimony that he did not know he had damaged Haynes and thus could not be liable for fraud despite the fact that two other qualified physicians swore that Smith had to have seen the damage if he did the procedure the way he claimed and if his eyes were not closed. The Court further rejected the challenge to the mother's consent and the claim that circumcision is neither medical nor surgical treatment as defined by law. Rather than discussing the latter argument in detail, the Court merely noted: "Pretermitting whether circumcision is medically necessary, it was a common, lawful surgical procedure performed in Georgia when Haynes was circumcised." Thus, the Court upheld the social norm and the pocketbooks of all Georgia doctors who circumcise. It declined to publish its opinion, thus eliminating any precedential value the case may have had and effectively hiding it from everyone but the litigants. Haynes petitioned the Georgia Supreme Court for review by certiorari, but that Court declined to grant the writ.

The Haynes case had some unfortunate racial and psycho-sexual overtones, which serve to illustrate the insanity of the social norm in favor of circumcision. The defendant physician testified that he had been non-circumcised until he was 25. At his deposition, when confronted with the proposition that circumcision is a disfiguring procedure, he stated: "Well, I don't agree with that. I think my penis looks a whole lot better now than it did before I was circumcised." The plaintiff was of black African ancestry and the defendant was white. When presented with a line drawing, showing an intact adult penis with a short acroposthion and, when asked to agree that it showed an uncircumcised or intact adult penis, he responded: "A lot of the penises that we circumcise, particularly in black folks, the prepuce was about twice that long." He added upon further inquiry that he was speaking of infants' penises rather than adult ones. However, his unexpected remark was reminiscent of the polemic of Peter C. Remondino, MD, who, in 1894, advocated universal circumcision of African-Americans, then commonly called Negroes and, in so doing, contended they were "a race proverbial for the leathery consistency, inordinate redundancy, generous sebaceousness, and general mental suggestiveness and hypnotizing influence of an unnecessary and rape, murder, and lynching breeding prepuce."[28]

Of course, these cases do not mean that all circumcision cases are futile. Far from it. They merely illustrate that, so long as the social norm confers "good parent" status upon those who circumcise, those who oppose circumcision will not find a "magic bullet" in the courts. Brave new theories of recovery have little chance of success unless the social norm changes. If a circumcision case is to be successful, it must be based on firm precedent and ancient tort principles. For example, boys who are circumcised without the consent of either parent have a fairly easy time collecting damages, although the amount may not be as great as the damage warrants. Likewise, men who have not given consent to a circumcision but who are circumcised anyway during other surgery have a remedy that generally, although not always, will be honored. Verdicts and settlements in the middle five-figure range are common in such cases. Boys who lose more than their foreskin in a circumcision can expect to recover substantial damages. Men whose erections are foreshortened by the physician's ignorance of normal anatomy or by their poor surgical technique also have the very real possibility of legal redress,[29] although those cases often hinge upon finding the right expert who understands normal anatomy and correct surgical technique. This can be somewhat of a challenge in the United States.[30] Even in cases where divorced parents are fighting over the circumcision of an older boy, a successful result of non-circumcision can result, but it requires a court willing to put the best interests of the child ahead of all other considerations and willing to seriously consider the medical evidence given by a qualified expert. Such experts are a rarity in the U.S.

But as Waldeck and Miller surmise, the tide may be turning and the social norm may be changing. The ostrich-like approach of sex education books finally has been broken with the publication of the third edition of the extremely popular *The What's Happening to My Body Book for Boys: A Growing Up Guide for Parents and Sons.*[31] Chapter 3 should be required reading, not only for boys and their parents, but also for all physicians and nurses in the United States. It portrays the natural, intact penis as normal, explains and illustrates the normal progression of foreskin retraction as a boy matures, explains and illustrates the importance of and function of the foreskin and its *ridged band,* identified by Dr. John Taylor,[32] and gives advice on "Hanging on to Your Foreskin." Best of all, it accomplishes this without leaving circumcised boys feeling devastated. This book, which may be the most widely read sex education book for boys in the United States, will help tremendously in shifting the social norm to non-circumcision in the next few years. Further change cannot help but come about as a result of the recent publication of *What Your Doctor May Not Tell You About Circumcision* by Paul Fleiss, MD, and Frederick M. Hodges, DPhil.[33] Published by Warner Books, this book attacks the social and medical myths that have grown up around circumcision.[34] Its wide distribution by a major United States publisher cannot help but have a good effect. Even less family-oriented publications are having a positive influence. For example, a Playboy playmate recently extolled the sexual pleasure provided to the female by the foreskin and stated that she would never circumcise her sons because to do so would rob their future wives of sexual pleasure.[35] Further, while agitation by anti-circumcision opponents may be irritating to some, it is necessary to raise public awareness. However, in the future it may be more profitable to take the general approach that a parent is a good parent and a smart parent for not circumcising, rather than suggesting that one is a bad parent for circumcising. Moving the social norm to non-circumcision is the object and that is best accomplished by positive reinforcement rather than negative polemics.

Can such a strategy work? I believe it can and already is. For example, in other English-speaking countries, the rates of circumcision are in serious decline. The College of Physicians and Surgeons in two Canadian provinces have essentially spoken out against neonatal circumcision and have warned physicians of potential legal consequences if they continue to perform unwarranted surgery on infants.[36] As the social norm shifts entirely away from circumcision in Canada, the effect will be felt in the United States. Only upon the rarest occasions do Americans like to feel isolated from the mainstream. Once the norm shifts in Canada and more circumcision-related suits are brought there, the United States courts will become more receptive to them also. Further, as the paradigm shifts in the United States and as more and more middle and upper class parents eschew circumcision, the courts will become more open to sustaining new legal theories. For example, in its decision in Haynes v. Smith, the Court of Appeals rested its rejection of the

[25] Private letter from John A. "Drew" Pumphrey, M.D., Associates in Urology, Fort Worth, Texas to Camille Azar, dated August 2, 2004.

[26] *Haynes v. Smith*, Georgia Court of Appeals, No. A04A1502. The entire opinion of the Court is appended hereto in Appendix A.

[27] Summary judgment ends a case as a matter of law on undisputed facts of record and removes the case from consideration by a jury.

[28] Remondino, P. Questions of the day:. Negro rapes and their social problems. National Popular Review. 1894; 4:3-6.

[29] Charles Bonner, Esq. of Sausalito, California and I recovered $501,550 last July for a Los Angeles couple in an arbitration against Kaiser Foundation Hospitals, *et al.*, in which it was contended that the previously circumcised husband was subjected to another "circumcision" with consequent loss of much of his shaft skin and significant reduction of the length of his erection.

[30] Most U.S. physicians seem unacquainted with the looseness of the penile skin system in the intact male and with the action of the Dartos fascia/muscle in the foreskin, which compresses the penis to protect it. Failure to push down on the public fat pad to determine penile shaft length and skin coverage before delineating the area of the outer incision during a sleeve circumcision seems to be one of the most common errors leading to excessive shaft skin removal and subsequent reduction in erection when the abdominal skin is then sewn to the mucosa behind the glans. Perhaps if we had more ancient or Renaissance statutes on display in the U. S. this would be less of a problem. As it is, this error appears to be quite common.

[31] Madaras L. Madaras, A. The what's happening to my body? Book for boys: A growing up guide for parents and sons. New York, N; Newmarket Press:2000.

[32] Taylor J., Lockwood A., Taylor A., The prepuce: specialized mucosa of the penis and its loss to circumcision. B J Urol 1996; 77:291-5.

[33] What your doctor may not tell you about CIRCUMCISION: Untold facts on America's most widely performed - and most unnecessary - surgery. New York, NY: Warner Books; 2002.

[34] Unfortunately, it has come under attack by one of circumcision's leading proponents. The authors have filed suit against Thomas E. Wiswell, M.D. for allegedly libeling them in a review of the book at www.amazon.com, in which Wiswell allegedly stated that "their 'facts' are untold because they are lies and diatribe," and that "what these two individuals put forth is as far from the truth as any author can get," and in which he is alleged to have made other defamatory statements about them. The defendant filed a motion to dismiss on the ground that the alleged defamatory statements were legally protected opinion, which motion the plaintiffs opposed, but the court granted. The case is now on appeal to the U. S. Court of Appeals for the Second Circuit. *Fleiss v. Wiswell,* C. A. No. CV 040964, U.S.D.C., E.D.N.Y.; *Fleiss v. Wiswell,* Docket No. 05-0161-cv, U.S. Ct. App. 2d Cir.

[35] http://www.playboy.com/sex/cos/heidimark/, accessed July 7, 2004.

[36] College of Physicians and Surgeons of Saskatchewan. Caution against Circumcision of Newborn Male Infants. Feb 2002; College of Physicians and Surgeons of British Columbia. Policy Manual. Infant Male Circumcision. Jun 2004, http://www.cpsbc.bc.ca/policymanual/c/c13.htm.

[37] Undoubtedly with a religious exception carved out to protect Jewish and Muslim parents.

Chapter 20

OF WASTE AND WANT

A Nationwide Survey of Medicaid Funding for Medically Unnecessary, Non-Therapeutic Circumcision

Amber Craig, Dan Bollinger
International Coalition for Genital Integrity

Abstract: The United States Medicaid Program funds over one-fourth of the 1.2 million unnecessary infant circumcisions performed each year at a cost exceeding $145 million, including doctor's fee, circumcision repair, and extended hospital stay. Additional lifetime costs bring the annual total to $443 million, about $11.6 million per state. The incidence of infant circumcision has been steadily declining over the past twenty-five years, from a time where almost all boys were circumcised to the present day, when about half are. During that same time period, costs have risen dramatically. With the exception of the United States, those few countries that adopted mass circumcision have already abandoned the practice. Medicaid budget shortfalls demand that only medically valid procedures be funded, and funding for cultural and cosmetic niceties be cut. Sixteen states do not fund infant circumcision, more are discussing defunding, and many private insurers do not cover infant circumcision.

1. INTRODUCTION

Infant circumcision entails the excision of healthy, normal, functional tissue from the penis of a newborn baby boy. United States taxpayers fund approximately twenty-eight percent of non-therapeutic infant circumcisions performed in this country.[1] The American Medical Association (AMA), the American Academy of Pediatrics (AAP), the American Academy of Family Physicians (AAFP), and the American College of Obstetricians and Gynecologists (ACOG) consider circumcision of male infants to be a

G.C. Denniston et al. (eds.), Bodily Integrity and the Politics of Circumcision, 233—246.

medically unnecessary procedure performed for cosmetic, social, or religious reasons. These organizations do not recommend routine infant circumcision.

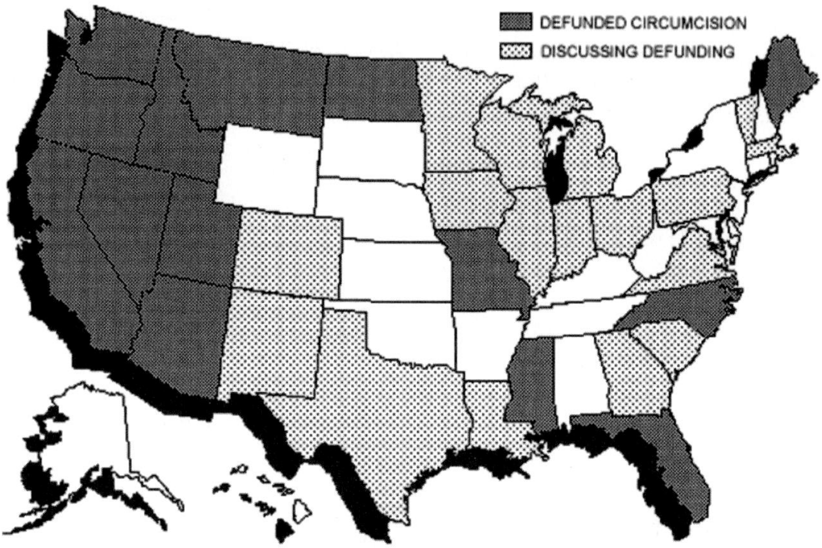

The state Medicaid programs in Arizona, California, Florida, Idaho, Maine, Mississippi, Missouri, Montana, Nevada, North Carolina, North Dakota, Oregon, Utah, and Washington have policies against funding circumcision except for a demonstrated medical need, saving taxpayers millions of dollars.

2. CURRENT PURPORTED BENEFITS OF CIRCUMCISION

The potential benefits suggested by circumcision proponents today include a slight decrease in urinary tract infections (UTI) for babies under one year, a claimed reduction in the risk of contracting sexually transmitted diseases, and the prevention of penile cancer. The AMA, in its 1999 position paper, emphasized that these potential benefits are insignificant and cannot justify prophylactic removal of the foreskin or the risks and possible complications that result from prophylactic circumcision.[2]

Chessare found non-circumcision to produce the highest medical utility.[3] In 1999, the AAP, after reviewing over forty years of research data on the potential benefits of circumcision, issued a policy statement concluding, "these data are not sufficient to recommend routine neonatal circumcision." [4]

3. CURRENT CIRCUMCISION PRACTICE IN THE UNITED STATES

Currently, fifty-nine percent of newborn males are circumcised annually in the United States.[5] Circumcision rates vary widely by region. In 1997, the Midwest had the highest rate of neonatal circumcision, 81.6 percent, while the West had the lowest rate, thirty-eight percent.

4. SURVEY RESULTS

4.1 Monetary Expenditures

Thirty-eight states have not defunded circumcision at the writing of this report (Table I). Circumcision costs are on the increase. On average, each state spent $751,378 annually on circumcision in 2003, seventy-seven percent more than the 1999 average national expenditure. Allowable reimbursement fees increased $11 for code 54150, on average. Of the thirteen states that reported detailed information for both the 1999 and 2003 reports, $10,818,504 was spent on circumcision in 2003, up thirty-seven percent for these states from $7,907,824 in 1999 (Table II).

4.2 Medicaid Circumcision Rates Down — Less Than Half of Medicaid Parents Choose Circumcision

Of the twenty-two responding states, their average infant circumcision rate was 56.1 percent, below the national average. Comparing the thirteen dual-reporting respondents' infant circumcision rates to previous years, our survey found that the infant circumcision percentage rates had increased 4.6 percent on average, while the number of births covered in those states increased twelve percent. Statistically, fewer Medicaid recipients are requesting circumcision.

TABLE I - CIRCUMCISIONS COVERED BY STATE MEDICAID PROGRAMS - SFY 2003

	Male	Infants Circ'd		Circumcisions Performed					Reimbursement Fees					Medicaid Reimbursements					State
	Births	Number	Percent	54150	54152	54160	54161	54163	54150	54152	54160	54161	54163	54150	54152	54160	54161	54163	Expense
Alabama	7783	4832	62.1%	4575	94	257	9	21	114	147	101	142		507452	13467	32575	1460	2679	$557.623
Alaska		1470		1456	10	14	136	5						426032	3435	4950	32944	930	$468.291
Arkansas																			
Colorado		5676																	$272.941
Connecticut	5859	2285	39.0%					24	69	86	87	122						3710	$298,967
Delaware	5728								113	140	143	199							
Dist. of Col.		101		96	3	5	22		41	80	151			3935	132	400	5139		$9,606
Georgia		9378		18,820	327	1,439	1,466		155	133	199	213		1,463,563	23,210	193,198	342,349		$2,012,320
Hawaii		224		219	0	5	7		85	120	129	157		18604	0	425	595		$19.623
Idaho	4248	2981	70.2%	2956	4	25	28	9	125	139	250	200		368730	454	2833	5389	1596	$379.002
Illinois	18552	14624	78.3%	13481	196	1043	2701	71	78	140	100	147		1017178	24446	98585	346505	8829	$1.486.713
Indiana		8934		8458	50	476	584	28						525204	4673	52466	58430	2851	$694.624
Iowa	5630	3604	64.0%	3572	17	32	115							533224	13532	8574	111343		$666,673
Kansas									56	138	93	177	34						
Kentucky	16624	7270	43.7%	6879	53	391	52		68	120	121	166		464319	5518	46974	7190		$524,001
Louisiana	20638	12917	62.6%	12325	26	592	16	46	81	135	72	204	192	996954	3607	40674	2697	9424	$1,053,356
Maryland	1034	445	43.0%	399	0	46	1	5	18	42	24	59		7333	0	1114	59	551	$9.057
Massachusetts									101	115	190	150	156						
Michigan	15338	8914	58.1%	8690	39	224	130		118	95	172	125		1022847	3036	21920	14611		$1.062.314
Minnesota*	12378	8730	70.5%	8652	203	78	980		55	69	53	193		475860	14007	4134	189141		$683.141
Nebraska	5134	3064	59.7%	3044	19	20	32	17	84	50	50	199		256822	943	993	6353	5638	$270,748
New Hampshire	3814	1065	27.9%	1031	6	34	72		80	80	80	83		82488	494	2734	5928		$91,644
New Jersey																			
New Mexico																			
New York																			
Ohio	23408	20965	89.6%	20045	897	920	3046	130						1032727	82433	70850	363186	18592	$1.567.788
Oklahoma																			
Pennsylvania	24272	10498	43.3%	10171	75	327	863	25	79	59	128	128	148	807328	7215	40428	116271	12736	$983.978
Rhode Island																			
South Carolina	18509	11996	64.8%	11535	72	461	147	127	65	110	115	150		753943	7981	50286	17868	14155	$844,233
South Dakota																			
Tennessee																			
Texas	97169	25805	26.6%	23107	958	2698	5644	125	48	181	91	182		1102047	173158	244576	1025047	19503	$2.564.331
Vermont																			
Virginia									190	106	195	141							
West Virginia																			
Wisconsin																			
Wyoming									115	139	91	187							

TABLE II – STATE COMPARISONS FOR 1999 AND 2003

	Male Live Births			Number of Circumcisions			Circumcision Rate			54150 Code Fee			Est. Expense*		
	1999	2003	Change	1999	2003	Change	1999	2003	Change	1999	2003	Change	1999	2003	Change
Alabama	15,000	7783	-48%	9065	4832	-47%	60.4%	62.1%	3%	$114	$114	0%	$1,237,226	$557,623	-55%
Connecticut	5,728	5859	2%	2,101	2285	9%	36.7%	39.0%	6%	$44	$69	57%	$93,002	$298,967	221%
Idaho	3,000	4248	42%	2,323	2981	28%	77.0%	70.2%	-9%	$94	$125	33%	$219,748	$379,002	72%
Illinois	15,522	18562	20%	12,174	14524	19%	79.4%	78.3%	-1%	$60	$78	30%	$1.191.127	$1.486.713	25%
Kentucky	15,214	16624	9%	5,022	7270	45%	33.0%	43.7%	32%	$68	$68	0%	$415.738	$524.001	26%
Louisiana	21,116	20638	-2%	10,763	12917	20%	51.0%	62.6%	23%	$81	$81	0%	$1.213.704	$1.473.809	21%
Michigan	12,968	15338	18%	6,842	8914	30%	52.8%	58.1%	10%	$50	$87	73%	$376.857	$1.062.314	182%
Minnesota*	12,400	12378	0%	5,423	8730	61%	43.7	70.5%	61%	$55	$55	0%	$369.399	$683.141	85%
Nebraska	4,382	5134	17%	2,519	3064	22%	57.5%	59.7%	4%	$37	$84	124%	$105.937	$270.748	156%
New Hampshire	2,055	3814	86%	709	1065	50%	34.5%	27.9%	-19%	$80	$80	0%	$57.176	$91.644	60%
Pennsylvania	8,890	24272	173%	4,305	10498	144%	48.4%	43.3%	-11%	$79	$79	0%	$463.630	$983.978	117%
South Carolina	14,617	18509	27%	10,105	11996	19%	69.1%	64.8%	-6%	$65	$65	0%	$780,948	$844,233	8%
Texas	91,540	97169	6%	19,562	25805	32%	21.4%	26.6%	24%	$51	$48	-5%	$1,391,332	$2,564,331	84%
TOTAL	224,431	252,321	12%	95,946	116,884	22%							$7,907,824	$10,818,504	42%
AVERAGES	17110	19255	26.84%	7227	8837	28.80%	53.02%	54.36%	4.59%	68	79	24.08%	608140	832039	77%

4.3 Code 54163 —Circumcision Repairs

For the thirteen states that responded with information on our question about circumcision repairs, we found 633 procedures, billed at an average reimbursement of $133, for a total of $101,194. According to data collected, 0.57 percent of circumcised neonates require a circumcision repair, making the complication rate from circumcision substantially higher than the .02 to .06 percent currently cited by the AAP. Extrapolating to the 38 states (and the District of Columbia) that cover circumcision, the nationwide total number of circumcision repairs paid for by Medicaid is about 1,904 procedures for a total cost of $253,232.

4.4 Florida Defunds Circumcision

Beginning 1 July 2003, Florida no longer funded Medicaid non-therapeutic circumcision. In the fiscal year prior to defunding (1 July 2002 through 30 June 2003), Florida Medicaid paid for 13,749 children under the age of one to be circumcised, at a taxpayer cost of $2,240,166. For fiscal year 2003-2004, Florida Medicaid paid for 1,056 children under age one to be circumcised, at a cost of $169,784, a direct savings of over $2 million.

5. DISCUSSION

Numerous states offered cultural and social justifications for covering the procedure; however, physicians are not ethically or legally permitted to serve as cultural brokers. Medicaid agency complacency toward the utilization of urgently needed health-care dollars for a harmful, medically unnecessary procedure is somewhat remarkable. A laudable interest in promoting cultural heritage cannot be permitted to mandate the diversion of funds earmarked for crucially needed medical care to cover a cultural procedure that has no medical justification. Given current knowledge, Medicaid dollars should not be used to fund circumcision any more than they should be used for other societal and cultural practices, such as body piercing, tattoos, female circumcision, or foot binding.

5.1 Scarce Medicaid Dollars Funding 28 Percent of All Medically Unnecessary Circumcisions

In the United States, 1.2 million newborn males are circumcised annually. Taxpayers are funding twenty-eight percent of medically unnecessary newborn circumcisions through the Medicaid program.[6]

5.2 Hidden Cost: Hospital Stay Quadruples Medicaid Expense

Professor Christopher J. Mansfield of the East Carolina University School of Medicine, lead author of a cost-factor study that includes the length of hospital stay associated with neonatal male circumcision, reports that, when a male infant is scheduled for circumcision, both mother and child remain in the hospital 0.26 day (about six hours) longer than they would without circumcision.[7] Mansfield, *et al.,* report that, during that stay, both mother and baby consume large and expensive hospital services, including room, board, nursing services, nursery care, and other services. Mansfield, *et al.,* estimate that the longer stay by each mother and child is billed at $900 per day, or $234 per circumcised child (1990-1991 dollars).

According to Bureau of Labor Statistics data on the Consumer Price Index, adjustment for inflation raises this amount forty-seven percent giving an adjusted cost of $344 per circumcision (2003-2004 dollars). Considering the estimated 336,152 Medicaid-reimbursed circumcisions performed in 2003, the total direct reimbursements of $29.6 million is increased $116 million for the additional hospital stay due to the procedure. The additional cost for hospital stay increases Medicaid reimbursements for circumcision 3.9 times. Thus, without even accounting for a range of additional cost factors — diagnostically related group or facility fee; anesthesia/analgesia; supplies; non-physician staff salaries; and treatment of complications — the total cost for Medicaid circumcision reimbursements exceeds $145 million nationwide, about $3.8 million per state, and this is not the total bill.

Table III. Estimated Medicaid Expenses for Circumcision

	Number	Cost	Total
Average Doctor's Fee	336,152	$88	$29,581,376
Extra 0.6 Day Hospital Stay	336,152	$344	$115,636,288
Circumcision Repairs	1904	$133	$253,232
Additional Lifetime Costs	336,152	$828	$278,333,856
Total Estimated Expenses	–	–	$443,203,712
Estimated Expenses Per State	–	–	$11,663,256

A 2004 cost-utility study of male neonatal circumcision was published shortly before completion of this report. That study, authored by Robert S. Van Howe, MD, MS, FAAP, of the Michigan School of Human Medicine, found that, for every 1000 circumcisions, 15.3 man-years of healthy life are lost, along with a cost increase of $828.42 per patient. This yields an estimated additional lifetime cost due to infant circumcision to be $278 million, should Medicaid cover these men at that time.[8] Van Howe reports that neonatal circumcision causes a net decrease in health and cannot be justified medically or financially. The ill effects of male neonatal, non-therapeutic circumcision reported by Van Howe are even greater than had generally been believed. This new study effectively obsoletes all previous medical association statements. Van Howe asserts:

> If neonatal circumcision was cost-free, pain-free, and had no immediate complications, it was still more costly than not circumcising. Neonatal circumcision is not good health policy, and support for it as a medical procedure cannot be justified financially or medically.

All told, this brings the total estimated Medicaid expenditures due to infant circumcision to be $443 million, or about $11.6 million per state, per year.

5.3 Federal Law Prohibits Payment for Infant Circumcision using Federal Funds

Medicaid is a creation of Federal law. Medicaid law, for the most part, is codified in the United States Code at Title 42 (Public Health and Welfare), Chapter 7 (Social Security), Subchapter XIX (Grants to States for Medical Assistance Programs).

Congress provides grants to states to furnish "necessary medical services." Congress has charged the Centers for Medicare and Medicaid Services (CMS) with determining which services are medically necessary. The CMS has determined that certain Ninth International Classification of Diseases, Clinical Modification (ICD9-CM) codes do not support medical necessity. Among those codes is ICD9-CM V50.2, circumcision at patient (or parental) request. This is the non-therapeutic circumcision without medical indication that is still being performed in the United States. Given this finding by the CMS, that circumcision at parental request is medically unnecessary, the authority to dispense federal funds to service providers for ICD9-CM V50.2, medically unnecessary non-therapeutic circumcision, is unclear. Such use of federal funds may be unlawful.

6. HOW MEDICAID FUNDING OF CIRCUMCISION HURTS THE POOR

Some advocates for the poor have argued that eliminating Medicaid coverage for elective non-therapeutic circumcision will take away options for poor parents. In light of the massive budget crisis most Medicaid agencies are experiencing and the numerous valid needs that are unmet, it is harmful to the poor to allocate money for an unnecessary cosmetic procedure while other genuine needs go unmet. Maintenance and expansion of medically justified, needed services for the poor are a high priority. In order to maximize the utility of scarce health-care dollars, government funds should not be used to pay for procedures based on parents' cosmetic preferences or religious choices. Many Medicaid directors cite, as the primary reason for continued coverage, the belief that the circumcision decision should be left up to the parent; therefore, they will reimburse. This is a false and inappropriate justification. Discontinuing Medicaid funding of circumcision does not prevent the parent from obtaining one if they have a strong cultural or religious desire to do so.

6.1 Medicaid's Promotion of Medically Unnecessary Treatment

According to the Centers for Medicare and Medicaid Services, a medically indicated circumcision requires a patient complaint, a diagnosis of pathology or physical abnormality, and conservative effective treatment for the diagnosed condition. Routine circumcision of the newborn does not meet the criteria for a medically necessary surgery because there is no documented pathology, physical abnormality, or complaint on the part of the patient. Therefore, routine circumcision is non-therapeutic. In its latest statement, the American Medical Association says, "The term 'non-therapeutic' is synonymous with elective circumcisions that are still commonly performed on newborn males in the United States."[9]

A study published in *Pediatrics* concluded that physicians significantly under-inform parents regarding the risks and asserted benefits of routine non-therapeutic circumcision.[10] This study found that nine out of ten parents were not given adequate information, and some male infants are being circumcised without parental consent or knowledge. This is especially true for economically or educationally disadvantaged parents.

The American Academy of Pediatrics stresses, in their current 1999 policy statement, that "Physicians counseling families concerning this decision should assist parents by explaining potential benefits and risks, and by ensuring that they understand circumcision is an elective procedure."[11]

Circumcision is perpetuated by the lack of accurate information offered to parents by medical professionals, as well as by the tacit approval Medicaid and some private insurance companies give to routine infant circumcision by unquestioningly funding this elective, medically non-indicated, non-therapeutic procedure. In its policy statement, the AAP offers parents a warning to avoid physician manipulation of them into assenting to the procedure: "Parents should not be coerced by medical professionals to make this choice." Unfortunately, many Medicaid recipients may interpret government reimbursement for circumcision as an official stamp of approval for circumcision. Just as Medicaid parents believe that covered medical services such as routine vaccinations and regular prenatal care are necessary and important for their child's health, it would be natural for them to conclude that, if Medicaid pays for circumcision, it must be medically beneficial and necessary for the health and well-being of their child.

6.2 Medicaid Promotes Infant Circumcision

In every reported instance where the Medicaid fee for infant circumcision increased from 1999 to 2003, so did the number of circumcisions performed in those states. A similar monetary connection was observed in the 1999 report. A Medicaid child born in a state that pays a physician a higher fee to circumcise is almost twice as likely to have his foreskin removed than a Medicaid child that lives in a state where the Medicaid reimbursement for circumcision is low. Medicaid-covered births increased twenty-seven percent from 1999 to 2000, but circumcisions covered increased twenty-nine percent, yielding a 4.6 percent overall circumcision rate increase, following a twenty-four percent average fee increase (Table II).

6.3 States Cut Medically Needed and Beneficial Services, While Paying for Elective Circumcisions

Budget shortfalls in many states are forcing Medicaid programs to consider the elimination of coverage for important, medically beneficial services. A report released in October 2004 by the Kaiser Commission on Medicaid and the Uninsured found that state Medicaid budget projections for 2004 indicated budget shortfalls of around $70 billion.[12] Every Medicaid program in the US initiated some type of cost containment measure during fiscal year 2003. It is surprising that any Medicaid agency would continue to expend desperately needed healthcare dollars on any medically unnecessary and controversial procedure —be it circumcision or any other cosmetically desired procedure that is not medically beneficial for the patient. Yet, thirty-eight states and the District of Columbia continue to do just that —allocate scarce healthcare dollars to fund elective circumcisions if parents want them, while cutting funding for medically indicated services. The elimination of coverage for circumcision frees up Medicaid funds that can be applied to meet children's genuine medical needs. For example, in fiscal year 2003, 513,000 covered citizens lost funding for prosthetics, orthodontics, eyeglass coverage, chiropractic services, and dentures. Apparently, not having teeth with which to eat or a replacement limb are considered less of a priority than ensuring amputation of a healthy, functioning part of an infant's penis. In Connecticut, funding was cut for speech, physical and occupational therapy, as well as psychological services, yet reimbursement for elective circumcision remained intact. In addition to Connecticut and Massachusetts, Georgia, Michigan, Nebraska, New Hampshire, Ohio, Pennsylvania, Texas, and Vermont also took away medically justifiable benefits such as dental care, podiatry, eyeglasses during fiscal year 2003, or had plans to remove coverage in fiscal year 2004, while continuing to fund medically unnecessary circumcision. Just prior to publication, Louisiana and Minnesota reported that they will no longer cover infant circumcision.

6.4 Increasing the Number of Uninsured, While Paying for Cosmetic Circumcisions

Some of the most concerning cost-containment measures that state Medicaid programs have taken have been eligibility reductions. Hundreds of thousands of people, once covered under Medicaid, were forced off the program and into the ranks of the uninsured —while medically unnecessary, non-therapeutic circumcision remained on the list. During fiscal year 2003, Alaska, Arkansas, Colorado, Connecticut, Indiana, Iowa, Kansas, Kentucky, Louisiana, Massachusetts, Nebraska, Ohio, Oklahoma, Rhode Island, South

Carolina, and Tennessee all continued to pay for unnecessary circumcision, while trimming thousands of needy citizens from their roles. For fiscal year 2004, Texas changed eligibility guidelines for Medicaid, resulting in nearly 8,000 pregnant woman and 8,400 adults considered "medically needy" losing coverage, and an additional 28,000 other adults to be dropped as well. Connecticut changed its eligibility rules to eliminate coverage for thousands of poor adults. In addition, Alaska, Minnesota, and Nebraska also took measures for fiscal year 2004 that resulted in thousands of children previously covered to no longer have any insurance. Medicaid agencies are continuing to waste taxpayer money by funding medically unnecessary, unwarranted circumcision at the whim of a parent, while dropping innocent children from low-income families off the Medicaid rolls. Obviously, eliminating coverage for circumcision cannot and will not provide all the needed funding to cover those dropped from the rolls. However, a few million dollars, no longer wasted on elective circumcision, can and will help cover more low-income citizens.

7. CONCLUSION

Medically necessary circumcision is extraordinarily rare. Circumcision is performed, not for any medical requirement, but for social, religious, and /or cultural reasons, reasons that have nothing to do with the purposes to which scarce Medicaid funding is dedicated. In their position statements, both the AAP and the AMA highlight the fact that "a number of medical societies in the developed world have published statements that do not recommend routine circumcision of male newborns." Yet, Medicaid paid for 336,152 circumcisions in 2003.

States reported average payment of physician fees for circumcision of $88, which projects to a total $29.6 million spent by Medicaid in 2003 just on physician fees for circumcision. Additional cost factors include extended time spent in hospital ($116 million), circumcision repairs ($253,232), and lifetime healthcare costs due to general poorer health associated with being circumcised ($278 million). On average, states pay an estimated $11.6 million per year for infant circumcision, its costs and complications.

In addition, there are the unknown costs of managed care enrollment, facility fees, anesthesia, Diagnostically Related Group fees, supplies, non-physician staff salary, costs incurred due to medical complications, and legal liability for "botched" and unauthorized circumcisions. One of the ten most common reasons for which pediatricians are sued is botched circumcision.

Through Medicaid, the United States taxpayers are currently funding more than twenty-eight percent of all newborn circumcisions.[13] This waste of tax dollars continues despite recent policy statements by the American Academy of Pediatrics and the American Medical Association who do not recommend this procedure. Many American taxpayers are increasingly concerned that the US federal and state governments continue to fund this non-therapeutic elective procedure when currently many economically deprived children's healthcare needs go unmet. Parents who elect this medically unnecessary non-therapeutic surgery should assume the cost.[14]

Arizona, California, Florida, Idaho, Louisiana, Maine, Minnesota, Mississippi, Missouri, Montana, Nevada, North Carolina, North Dakota, Oregon, Utah, and Washington do not fund circumcisions, saving their state taxpayers millions of dollars. Many state Medicaid agencies and legislatures are discussing whether or not they will continue to provide coverage for routine, medically unnecessary, non-therapeutic circumcision. Most private insurers do not cover infant male circumcision.[15]

Regardless of the cultural or religious preference of the parent requesting this non-therapeutic, elective procedure, Medicaid should not utilize taxpayer dollars to fund procedures that are not recognized by the medical community as offering a significant health benefit. Many taxpayers oppose the use of public tax dollars to fund a non-medical procedure, especially when numerous critical children's healthcare needs must go unmet because of limited public healthcare funds. Tax dollars funding Medicaid programs should be directed to necessary medical services so desperately needed by all children in the United States and not wasted on medically unnecessary circumcision of male children. Desperately scarce Medicaid tax dollars should not be used to fund a cultural, religious, or ethnic tradition.

REFERENCES

[1] Bollinger D. Intact Versus Circumcised: Normal versus Circumcised: U.S. Neonatal Male Genital Ratio. Circumcision Reference Library (an original online publication), 22 April 2003. Available from: URL: http://www.cirp.org/library/statistics/bollinger2003/.

[2] Council on Scientific Affairs, Neonatal circumcision. Chicago: American Medical Association, 1999. Available at URL: http://www.ama-assn.org/ama/pub/article/2036-2511.html.

[3] Chessare JB. Circumcision: is the risk of urinary tract infection really the pivotal issue? Clin Pediatr (Phila). 1992 Feb;31(2):100-4.

[4] Task Force on Circumcision. Circumcision policy statement. American Academy of Pediatrics. Pediatrics. 1999 Mar;103(3):686-93.

[5] Bollinger D. Intact Versus Circumcised: Normal versus Circumcised: U.S. Neonatal Male Genital Ratio. Circumcision Reference Library (an original online publication), 22 April 2003. Available from: URL: http://www.cirp.org/library/statistics/bollinger2003/.

[6] Bollinger D. Intact Versus Circumcised: Normal versus Circumcised: U.S. Neonatal Male Genital Ratio. Circumcision Reference Library (an original online publication), 22 April 2003. Available from: URL: http://www.cirp.org/library/statistics/bollinger2003/.

[7] Mansfield CJ, Hueston WJ, Rudy M. Neonatal circumcision: associated factors and length of hospital stay. J Fam Pract. 1995 Oct;41(4):370-6.

[8] Van Howe RS. A cost-utility analysis of neonatal circumcision. Med Decis Making. 2004 Nov-Dec;24(6):584-601.

[9] Council on Scientific Affairs. Neonatal circumcision. Chicago: American Medical Association, 1999.

[10] Adler R, Ottaway S, Gould S. Circumcision: We have heard from the experts; Now let's hear from the parents. Pediatrics 2001;107, e20.

[11] Task Force on Circumcision. Circumcision policy statement. American Academy of Pediatrics. Pediatrics. 1999 Mar;103(3):686-93.

[12] Smith, Vernon, Rekha Ramesh, *et al.* The Continuing Medicaid Budget Challenge: State Medicaid Spending Growth and Cost Containment in Fiscal Years 2004 and 2005: Results From a 50 Survey. Kaiser Commission on Medicaid and the Uninsured. Kaiser Family Foundation, September 2003.

[13] Bollinger D. Intact Versus Circumcised: Normal versus Circumcised: U.S. Neonatal Male Genital Ratio. Circumcision Reference Library (an original online publication), 22 April 2003. Available from: URL: http://www.cirp.org/library/statistics/bollinger2003/.

[14] Cadman D, Gafni A, McNamee J. Newborn circumcision: an economic perspective. Can Med Assoc J 1984;131:1353-5.

[15] Desiree Hunter. Medical center policy affects circumcisions. The Advocate, Baton Rouge, 17 July 2003:5B.

Chapter 21

A CAMPAIGN FOR THE ERADICATION OF INFIBULATION WITHIN AN EXTENDED FAMILY

Khartoum, Sudan

Nagla Dawelbait, Pia Grassivaro Gallo, Marianna Pappalardo
Working Group on Female Genital Mutilations. Department of General Psychology, University of Padua (Italy)

Abstract: Between 1993 and 1995, a Sudanese extended family became the subject of an extraordinary local campaign regarding the issue of infibulation: Nagla is one member who, in response to the problem, began the difficult task of eradicating infibulation within her family. Through meetings and discussions, in which all family members took part, using educational materials made available by the Babiker Bedri organization, a final decision was reached: none of the family's young girls would henceforth undergo the ritual. Analytical evaluation and verification has been carried out in collaboration with the Padua Working Group on Female Genital Mutilation in the summer of 2004. As a result of this campaign, two hundred girls were spared infibulation. Some of them, however, are facing difficulties socializing with their peers. This paper presents an account of a successful attempt to eradicate this mutilating practice.

1. INTRODUCTION

Female genital infibulation is practiced in East Africa. The practice is found in the Sudan and other countries in the region. In the Sudan, female genital cutting (FGC) is inflicted on approximately ninety percent of women, especially in the northern part of the country (Aldareer, 1983; DHS, 1989-1991; Motherhood Survey, 1999-2001; Gruenbaum, 2004). Whereas, in the southern part of the Sudan, the practice tends to be less common.

Our area study focuses on central Sudan. In this region and, in particular, the eastern bank of the River Nile, epidemic diseases caused by FGC have

G.C. Denniston et al. (eds.), Bodily Integrity and the Politics of Circumcision, 247—260.

been surveyed. This research was sponsored by the Executive Secretary of the Sudan National Committee on Harmful Traditional Practices or SNCTP between 1996-2000. (Hassan, 2000; in press, 2001).

2. INFIBULATION AND PSYCHO-PHYSICAL MATURITY

Among all East African peoples, FGC is a sensitive issue within the community and the family. Discussions about FGC revolve around scheduling the event rather than debating its necessity.

Usually, it is the girl's mother who organizes the time, place, and hires the ritual operator. Otherwise, the opinions of all family members are considered, but the maternal grandmother's opinion is highly influential. The place where the operation takes place is rarely salubrious because it is performed under unsanitary conditions by a traditional operator or a midwife either in her own house or in the girl's house. In the past, the circumcision was a part of a midwife's job, but today the operation is no longer legally taught in the midwifery schools in order to discourage the practice. For this reason, the operation entails considerable risk because it is invariably performed by an elderly midwife or by a young, untrained midwife. This is obviously a transitional period for infibulation.

In preparation for the operation, the girl's house is readied. The room where the girl will recover is given special attention. The room will be furnished with ritual objects such as lamps, candlesticks, incense, and perfume, which are acquired for this occasion. The girl is dressed with particularly expensive clothes and covered by a colorful cloth (*firca*). Her hands and feet are decorated with henna and adorned with gold jewelry. The preparations are similar to those of a young bride on her wedding day. Photographs may be taken of the young girl in her finery.

After the infibulation, there is a party as on the wedding day. The girl receives visits from friends and relatives who offer gifts of money, jewels, and congratulations. She feels elated. Because a woman's birthday is not celebrated, the only occasions of personal rejoicing are those for circumcision, marriage, and childbirth. Following the operation, the girl's legs will be tied together. If she has undergone *sunna*, her legs will not be tied as tightly together. She is expected to lie still for approximately a month in her specially decorated room.

There is a current trend to celebrate the infibulation day more privately. The family does not invite all the neighbors to the party. Being torn away, without warning, from her childhood pursuits and her peers, the girl for the first time has the opportunity to observe closely a woman's life. She becomes the object of continuous attention by the women of the household who assist her during her convalescence. They provide her with proper food, help her to urinate, and monitor the progress of her healing.

After the seclusion period, the girl does not return to the life of a child, but she is now under the constant control of the women of the house. They teach her how a proper woman speaks and comports herself. If the wound tears apart, a reinfibulation may be necessary. After a few months, the infibulated girl achieves complete social and physical maturity and appears as a young lady "aware of all social aspects proper of womanhood like dress, speech, decency, which show she knows the separation between male and females and how to respect herself." (Hamid Al-Bashir Ibrahim, 2002). In particular, an infibulated young lady can be recognized from her gait, because spontaneous leg movement is limited out of fear that the scar might rip open.

2.1 Infibulation and Women's Childbearing Years

From the first sexual act on her wedding night to subsequent sexual intercourse with her husband, the protection of the small opening created through infibulation is a source of constant anxiety during a woman's childbearing years. The frequency of sexual intercourse can be reduced in order to avoid suffering risky and painful hole enlargement (Hamid Al-Bashir Ibrahim, 2002). By contrast, the natural, intact vulva is likened to a wide-open hole that might swallow both the penis and the testicles (Gruenbaum, 2001). In particular, the laboring woman will be socially either esteemed or dishonored according to the difficulty or facility of her childbirth, testified by her pain and suffering.

The midwife, therefore, is compensated for an operation with greater lacerations than needed, which permits a more severe reinfibulation, for the satisfaction of the relatives. A very difficult childbirth is usually the result of severe and tight infibulation. In this case, the midwife waits until the doctor leaves the room and then performs the operation.

Another type of vaginal seam, called *al-kabr* (tombs' infibulation), seems to exist. It is performed on an elderly woman at the end of her childbearing years on the occasion of a pilgrimage to Mecca, "to stop sexual intercourse with her husband, and to dedicate herself totally to prayers and other religious rituals, leaving earthly wealth and pleasure" (Hamid Al-Bashir Ibrahim, 2002; Sad Al-Fadil, in Lovel, *et al.*, 2004).

attenuation of the practice and the use of other types of *sunna* (Gruenbaum, 2000), with a more or less complete seam, similar to infibulation. Some Sudanese view less severe forms of female circumcision to be more in keeping with the *hadiths* of the prophet Mohammad's.

Consequently, certain social factors inhibit the eradication of female circumcision. These include cultural constructs about the female body, genital aesthetics, and beliefs about the sexual consequences of infibulation (Gruenbaum, 2004). There is room for flexibility in devising and implementing ideas for the eradication of infibulation.

3. THE CAMPAIGN FOR THE ERADICATION OF INFIBULATION WITHIN AN EXTENDED FAMILY: THE H. FAMILY

Nagla's extended family includes about six-hundred people. The patriarch was a man who moved from northern to central Sudan seven generations ago, counting from the generation of children at risk of infibulation. This man was married to four women from various clans. From these marriages, four distinct family units descended, one of which corresponds to Nagla's close family circle, in which the great-grandfather (of the F3) married two wives and is considered the head of the family. His first residence was in Omdurman, but later he moved to Medani, where he settled in a district that has the same name as the H. clan. Due to this man's entrepreneurship, a school was opened along with other social institutions that serve the community.

The family is constituted by a network of agricultural and commercial activities that have extended gradually outside the town of Medani to the surrounding villages. The group is characterized by very close social relationships because the new members from different clans or even from a foreign country always try to settle in the original neighborhood, which is still considered to be the clan's home base.

The heads of the other three family branches are cousins. Two are males and one is a female. All of them refer to Medani as their place of origin, even if they live elsewhere (such as in Khartoum) or migrated to other countries. Even today, the family is so close knit that, when a family celebration is organized, members all over the world try to attend.

Another reason for this strong cohesion is the endogamous marriage tradition. In particular, the women of the family marry their third cousins for their first wedding. Their women, therefore, do not marry out of the family. Because of this habit, many third generation women remain unmarried for lack of "acceptable" cousins. The rule is not equally strict for male members.

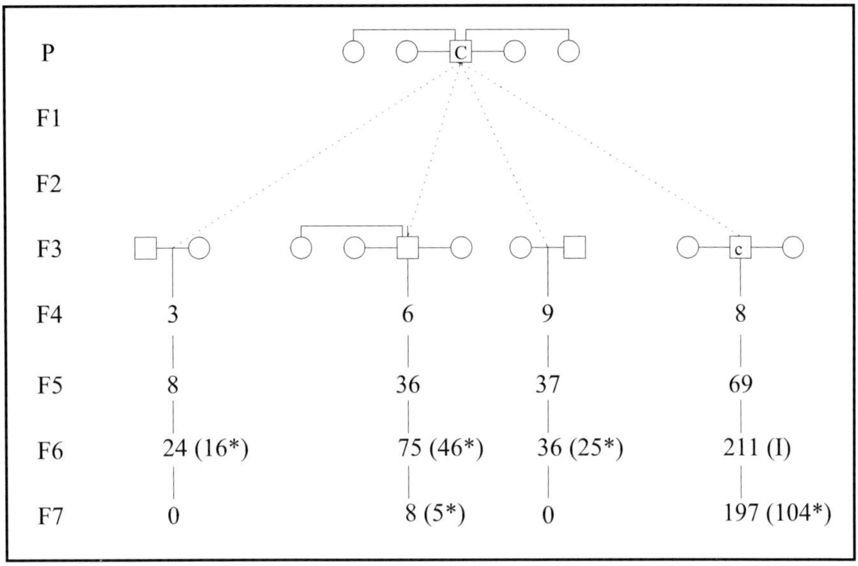

Figure 21-1. Family Tree.

□–○ = Single wife couple

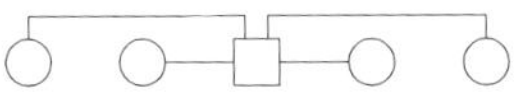

= polygenic couple

C c (I) = founder of the extended family; founder of N's family nucleus; information source (N.)

P = parent generation

F1.....F7 = offspring generation (information about F1and F2 are not included)

| = the 4 branches of the extended family

N (n*) = numeric values indicate the family members (male and female); girls at risk respectively

Finally, a further element of unity is religion: the *Quran* considers it a sin not to visit relatives and neighbors. The H. clan is formed by cousins who intermarry, have their residential core in the H. neighborhood of Medani, and recognize as their common founder a third-generation grandfather, counting from the generation of the children at risk. The other three parallel families developed from other cousins.

The kinship relationship among the members is consolidated in the traditional celebrations (weddings, births, infibulation, male circumcision, funerals, etc.). The clan has increased demographically through the acceptance of "foreign" women from other clans or countries who married with H. men. This is now commonplace since the family's agricultural and business activities are attractive to outsiders and because of emigration to study and work abroad. New members often integrate themselves better in the H. family than in the family into which they were born.

The children of the family belong to everybody, and all adults must care for them regardless of their kinship relationship. Adoption is difficult because orphans do not usually know their ancestry. Adopted children, however, have the same rights and duties as the biological offspring of the clan.

All decisions concerning female genital cutting are made by adults. The girl's opinion is never taken into consideration.

3.1 1993-1995: The Campaign

As mentioned above, family members gather periodically for celebrative occasions, such as births, weddings, and funerals, to strengthen their kin relations. At these gatherings, many topics are discussed. Nagla took the opportunity to raise the topic of harmful traditional practices. Along with the help of a group of cousins, Nagla promoted the campaign to end female circumcision within the H. family. Multiple individual meetings were held with both male and female relatives of all ages, excluding children.

For about two years, monthly meetings were held in both Khartoum and Medani, where most of the H. family live. These meetings targeted specific groups, such as men, women, and students, with the assistance of sociologists, psychologists, religious leaders, and doctors. The professionals belong to the Babiker Bedri organization, famous for the promotion of female schools and the establishment of the first university for women in the Sudan.

Houses of various relatives were used as meeting centers, and audiovisual aids and documents were supplied by Babiker Bedri.

For the final meeting, professionals met with the relatives who had participated in the meetings. They congregated in the garden of an aunt, who was the mother of six daughters and one son, to discuss infibulation.

The old women spoke first. These women had the strongest attachment to the tradition of infibulation and consequently had the greatest objections to abandoning the practice. Some fought to save infibulation while others acquiesced to the decision of the majority without being convinced.

The young mothers were unified in their desire to see infibulation eradicated, although one mother preferred to have only the *sunna* performed on her two children.

Next to speak were the young people who were at the age of marriage, traditionally with an infibulated partner. Many of them preferred to remain silent, but they were inspired by the impassioned words of a cousin who supported the deinfibulation of his fiancée in order to spare her the pain of the first sexual intercourse — the "breaking of the seam." At the height of the debate, another cousin proudly invited the entire family to abandon the practice.

It is important to remember that the campaign affected the entire family, including those members who were absent.

3.2 Summer 2004: Scientific Evaluation of the Campaign

After a few years, Nagla left Khartoum in order to continue her studies. News from her family arrived sporadically. In the summer of 2004 when Nagla returned to the Sudan, she contacted the Working Group on FGM at the University of Padua and asked for help in assessing the results of her campaign to eradicate FGM within her family. Two methods were used in the study:

1. Interviewing mothers of the "girls at risk" (those up to thirteen years old) to verify whether the promise made in 1995 to abandon female circumcision was upheld.

2. Estimating whether the 1993-1995 campaign contributed to changing the group members' opinion about female circumcision. A questionnaire was used for the mothers of the "girls at risk." The questionnaire would not only record information on the subject and her partner, but also on her ancestry, her mother, mother-in-law, their families, and her partner.

By September 2004, the qualitative data about the mothers and the possible social marginalization of the intact daughters had been obtained.

Moreover, another intensive questionnaire was issued to the mothers who had participated to the 1990s campaign.

3.3 Results

After this campaign, two-hundred girls in the H. family, under thirteen years of age, had been spared infibulation. They belong to all the four family branches, corresponding to the more recent generation (F6) (F7) in the family tree. In particular, one-hundred-four girls belong to Nagla's closest relatives. For the latter, it was possible to compare the data on fifty-one infibulated mothers with the two-thirds of mothers who were reinfibulated after delivery. The only exception was the daughters of one relative, four sisters whose father had decided secretly to leave them intact. We obtained the following answers to the question "What consequences have the non-infibulated daughters endured in their social intercourse and daily life?"

1. Twenty-four subjects had no problems. These were young girls with physical disabilities who attended school with other intact children, or who resided in a foreign country or in the Sudan for a short time.

2. Some difficulties were found for eleven intact girls, most of whom came from rural areas. One mother complained that her daughter was stressed by the continuous questions and pressures. Another specified that the problems worsened during the visits to the village.

3. Some mothers did not answer the question for the following reasons:

    a. Lack of communication between mother and daughter in eight cases.

    b. Two daughters were taught that they need not discuss the issue.

    c. Two subjects refused to answer the question.

In conclusion, our investigation verified that the campaign to eradicate infibulation in the H. family had been implemented. We also found that the campaign spread to neighboring families.

4. COMMENTS

Traditionally, infibulation has been an important tradition within Sudanese society, as with the Somali (Roheim, 1932). The entire society is affected by it. Women are affected during childhood, during childbearing years, and in old age when they undergo "tomb infibulation." Men are also affected by infibulation and hold strong views on the subject that often stand in stark contradiction to those traditionally held by women.

Another aspect that has been ignored by previous researchers is the effect of infibulation on the process of mental and physical maturation. After just a few months, girls who have undergone the practice are transformed into properly behaving Sudanese women. This important social aspect of infibulation should be taken into account in all campaigns to eradicate infibulation (see the proposal for the institution of pseudo-infibulation, Gruenbaum, 2004, witnessed in Somalia by Grassivaro Gallo, 1986).

The speedy transformation of infibulation in the Sudan has been hastened by a few creative individual campaigns in both rural and urban areas (Gruenbaum, 2001; 2004). These campaigns underscore an increasing awareness about the negative aspects of infibulation. It may still be very common, but there is a growing intolerance for the persistence of infibulation. In the eradication campaign that took place within an upper middle-class urban extended family, the instigators were a group of students at the University of Khartoum who belonged to the same family.

The results obtained were better than expected. Within a ten-year period and within a single generation, infibulation disappeared completely from the family core. There were many factors that contributed to the success of the campaign:

- The courage of these young people to propose innovative ideas in opposition to the traditional beliefs of the older generation whom they loved and respected.

- The patience of the young people in waiting two years to instigate an informative and educational dialogue among the members of the same community.

- The collaboration of professionals outside the family structure.

- The growing awareness among young Sudanese that infibulation is a harmful traditional practice without any value or necessity. They no longer attend parties in their neighbors' or relatives' houses to celebrate an infibulation.

Finally, there is the question of how intact girls coexist with their infibulated peers. Many mothers encourage their intact daughters to adopt the strategy of keeping silent about their condition. While this may make the intact girl's life easier, it does nothing to change public opinion. This is not the first time that a "conspiracy of silence" encourages the persistence of female genital cutting (Hosken, 1982).

5. CONCLUSION

Recently, Shweder (2003) has warned about the dangers of first-world "cultural globalization" and attempts to universalize Western culture, thereby violating human rights The global campaign against female genital cutting would seem to be a prime example of this kind of cultural imperialism. In response, we would point out that many campaigns to eradicate female genital cutting were initiated within African countries, independent of Western influence. This indicates that Africans themselves desire change. It is impossible to predict whether all of these campaigns will be successful. At present, anthropologists can serve the useful function of documenting the evolution of infibulation and the traditions and forces that perpetuate it, as well as offering explanations for the persistence of the practice (Gruenbaum, 1996).

Our work in East Africa has focused on *sunna gudnin* in Merka, Somalia (Grassivaro Gallo, *et al.*, 2001; 2004) and the Sudanese campaign described in this paper. Both campaigns had remarkable success although they differed in many ways. The first was a compromise in the form of a modified *sunna*, practiced with the same rites that formerly accompanied infibulation. The later campaign took the form of complete eradication of the surgical procedure as well as the cultural rites connected with it. We would like to emphasize some common elements in the two campaigns:

- They were created by African women.
- They did not involve Western or foreign people.
- They were non-violent, involving only conversation, persuasion, and education.
- Those involved in the campaign were all local people — religious leaders, traditional healers, midwives, doctors, etc.
- The campaigns aimed first at building consensus before making a declaration that infibulation would be abandoned. This was one

of the great strengths in the campaign for the eradication (or the suspension) of the practice (Makie, 1996; 2000).

In conclusion, we predict that scientific studies will verify the successful eradication of infibulation among new families formed by responsible intact women. Moreover, it is important to note that the Somali campaign preserved female genital cutting and its accompanying rituals while the Sudanese campaign eradicated the practice together with its associated cultural rituals.

BIBLIOGRAPHY

Abdel-Magied AM, El-Balah AS, Dawood MK Recircumcision: the hidden devil of FGM. Case study on the perception, attitudes and practices Sudanese woman. "Ahfad Journal," 17:22-32, 2000.

Ahmed S, Gruenbaum E, Abdel-Magied AM. Men break their silence on FGM on sexuality in Eastern Sudan. Seminar on Advancing. 2004.

Almroth L, Almroth-Berggren V, Bergstrom S, Hassanein OM, El Hadi N. Re-infibulation among women in a rural area in Central Sudan. "Health Care for Woman International," 22:711-721, 2001.

Bedri N. Impact of grandmothers on mother and child health, "Ahfad Journal," 12,1995.

El Dareer A. Epidemiology of Female Circumcision in the Sudan, "Tropical Doctor," 13: 41-45, 1983.

Gruenbaum E, The female circumcision controversy: an anthropological perspective. Philadelphia: University of Pennsylvania Press, 2001.

Gruenbaum E, Sexual response and Sudanese conceptions of the infibulated and uninfibulated vulva. A research in Sudan. Seminar on Advancing Knowledge of Psycho-sexual Effects of FGM/C: Assessing the Evidence, Alexandria, Oct.10-12, Egypt, 2004.

Hamid Al-Bashir Ibrahim. (original Arabic ed., 1997). La donna e la triste realtà culturale sudanese. Ricerca per capire l'ideologia dell'infibulazione delle ragazze in Sudan. (traduzione in italiano, www.islamitalia.it/donna/donnasudan.html, last revision 25/10/2002).

Hassan AR Amna. Resumè de l'impact de la récerche sur "Les MGF consequences psychosociales et sexuelles et changement d'attitudes dans KhartoumNord et Nile Est," Bull. du Comitè Inter-Africain, 28:5-7, 2000.

Hassan AR Amna. FGM historical background, views in Islamic shari'a et recent research's findings on FGM. (unpubl. data, 2001).

Hosken FP. The Hosken Report. Women's International Network News. 2nd Ed., Lexington, MA, 1982.

Lightfoot-Klein H. Prisoners of ritual: an odyssey into Female Genital Circumcision in Africa. New York, Harrington Park,1989.

Lovel H, Bedri N, Zeinab Mohamed, Njikam Y, Rhetta Moran, Mc Gettigan C, Efm Sa'ad, Khaled MA, Abdel Magied AM, Efua Dorkenoo. The practice of Re-infibulation, findings from a systematic search of English and Arabic literature. Seminar on Advancing Knowledge of Psycho-sexual Effects of FGM/C: Assessing the Evidence, Alexandria, Oct.10-12, Egypt, 2004.

Makie G. Ending footbinding and infibulation: a convention account. “American Sociological Review,” 61:999-1017, 1996.

Makie G. Female Genital Cutting: the beginning of the end. In: Shell Duncan B. & Hernlund Y. (Eds.) Female “Circumcision” in Africa. Culture, Controversy and Change, Lynn Rienner Publishers, Boulder (Co), 2000.

Roheim G. Psychoanalysis of primitive cultural types “X” The national character of the Somali. 1° Infibulation, “Internet. Journal of Psychoanalysis,”13:199-221, 1932.

APPENDIX

RESOURCES

1. ORGANIZATIONS

Association Contre la Mutilation des Enfants (A.M.E.). Didier Diers and Xavier Valle, Boite Postale 220, 92108 Boulogne Cedex, France. http://pages.pratique.fr/~ame1/.

Attorneys for the Rights of the Child, J. Steven Svoboda, JD, 2961 Ashby Avenue, Berkeley, CA 94705 USA. Tel: 510-595-5550. www. arclaw.org.

Circumcision Information Australia. www.circinfo.org/.

Circumcision Resource Center. Ronald Goldman, PhD. PO Box 232, Boston, Massachusetts, 02133 USA. Tel: 617-523-0088. www.circumcision.org/.

Doctors Opposing Circumcision (D.O.C.). George Denniston, MD, MPH. 2442 NW Market Street #42, Seattle, WA 98107 USA. Tel: 360-385-1882. Fax: 360-385-1965 http://weber.u.washington.edu/~gcd/DOC/.

Equality Now. Jessica Neuwirth, President. PO Box 20646, Columbus Circle Station, New York, NY 10023. Tel: 212-586-0906. Fax: 212-586-1611. www.forward.org.

Foundation for Women's Health Research and Development (FORWARD). Sarah Fisher, 765-767 Harrow Road, London NW10 5NY. Tel: 020-8960-4000. Fax: 020-8960-4014.

Inter-African Committee. Berhane Ros-Work, President. 147 rue de Lausanne, CH-1202 Geneva, Switzerland. Tel: 22-731-2420. Fax: 22-738-1823.

International Centre for Reproductive Health. Els Leye, FGM Project Coordinator. Ghent University, De Pintelaan 185 P3, 9000 Ghent, Belgium. Tel: +32-9 240.35.64. Fax: +32-9 240.38.67.

International Coalition for Genital Integrity. Dan Bollinger. Tel: 765-427-7012. www.icgi.org.

Israeli Association Against Genital Mutilation. Avshalom Zoossmann-Diskin, PO Box 56178, Tel-Aviv 61561 Israel. www.britmilah.org.

London Black Women's Health Action Project. Shamis Dirir. Cornwall Avenue Community Centre, First Floor, 1 Cornwall Avenue. London E2 0HW United Kingdom. Tel: 181-980-3503. Fax: 181-980-6314.

Medical Ethics Network. John Sawkey, PO Box 578, Yorkton, Saskatchewan, S3N 2W7. Tel: 306-744-2436. http://med-fraud.org.

National Organization of Circumcision Information Resource Centers (NOCIRC). [International Headquarters] Marilyn Fayre Milos, RN, Executive Dirrector. PO Box 2512, San Anselmo, CA 94979-2512. USA. Tel: 415-488-9883. Fax: 415-488-9660. http://www.nocirc. org/.

National Organization to Halt the Abuse and Routine Mutilation of Males (NOHARMM). Tim Hammond. www.noharmm.org/.

National Organization of Restoring Men (NORM). International Headquarters. R. Wayne Griffiths, MS, Med, 3505 Northwood Drive, Suite 209, Concord, CA 94520-4506 USA. Tel: 510-827-4066. Fax: 510-827-4119. //www.norm.org/.

NORM-UK. John P. Warren, MB. Chairman. PO Box 71. Stone, Staffordshire, ST15 0SF, United Kingdom. Tel/Fax: 01785-814-044. www.norm-uk.co.uk/.

Nurses for the Rights of the Child. Mary Conant, RN, Betty Katz Sperlich, RN, Mary-Rose Booker, RN. 369 Montezuma #354, Santa FE, New Mexico 87501. Tel: 505-989-7377. http://www.cirp.org/nrc/.

Rainb♀. Nahid Toubia, MD. 915 Broadway, Suite 1109, New York, NY, 10010-7108 USA. Tel: 212-477-3318. Fax: 212-477-4154.

Terres des Femmes. Petra Schnull, Gritt Richter, Claudia Piccolantonio. Kreuzbergring 10, D-37075 G ttingen, Germany.

2. WORLDWIDE WEB SITES

Alliance for Transforming the Lives of Children
www.atlc.org/

Association Contre la Mutilation des Enfants (French)
http://pages.pratique.fr/~ame1/

Attorneys for the Rights of the Child
www.arclaw.org/

Birth Psychology
www.birthpsychology.com/birthscene/circ.html

BoysToo.com (Official Website of NOCIRC of North Dakota)
www.boystoo.com

Circumcision Information and Resource Pages
www.cirp.org/

Circumcision Information Resource Center (Montreal, Canada)
www.infocirc.org/index-e.htm

Circumcision Resource Center (Boston, Massachusetts)
www.circumcision.org/

D.O.C. (Doctors Opposing Circumcision)
www.doctorsopposingcircumcision.org

Female Genital Mutilation Research Home Page
www.fgmnetwork.org/

In Memory of the Sexually Mutilated Child (John A. Erickson)
www.datasync.com/SexuallyMutilatedChild/

International Coalition for Genital Integrity
www.icgi.org/

Intersex Society of North America Home Page
www.isna.org/

Jews Against Circumcision
www.JewsAgainstCircumcision.org

National Organization of Circumcision Information Resource Centers
www.nocirc.org/

National Organization to Halt the Abuse and Routine Mutilation of Males
www.noharmm.org/

National Organization of Restoring Men (NORM)
www.norm.org/

NORM-UK (Great Britain)
www.norm-uk.org/

Students for Genital Integrity
www.studentsforgenitalintegrity.org/

3. BOOKS

Sami A. Aldeeb Abu-Sahlieh. *Male and Female Circumcision Among Jews, Christians and Muslims: religious debate.* Beirut, Lebanon: Riad El-Rayyes Books. 2000.

Sami A. Aldeeb Abu-Sahlieh. *Circoncision Masculine – Circonsion Femine: Debat Religieux, Medical, Social et Juridique.* Paris: L'Harmattan. 2001.

Jim Bigelow, PhD. *The Joy of Uncircumcising!* 2nd Edition. Aptos, CA: Hourglass. 1995. [ISBN 0-934061-22-x] www.norm.org.

Elizabeth Heger Boyle, *Female Genital Cutting: Cultural Conflict in the Global Community*. Baltimore. The Johns Hopkins University Press. 2002.

Robert Darby. *A Surgical Temptation: The Demonization of the Foreskin & the Rise of Circumcision in Britain.* Chicago. The University of Chicago Press. 2005.

George C. Denniston and Marilyn Fayre Milos, eds. *Sexual Mutilations: A Human Tragedy.* New York and London: Plenum Publishing Corporation. 1997.

George C. Denniston, Frederick Mansfield Hodges, and Marilyn Fayre Milos, eds. *Male and Female Circumcision: Medical, Legal, and Ethical Considerations in Pediatric Practice.* New York: Kluwer Academic/ Plenum Publishers.1999.

George C. Denniston, Frederick Mansfield Hodges, and Marilyn Fayre Milos, eds. *Understanding Circumcision: A Multi-Disciplinary Approach to a Multi-Dimensional Problem.* New York: Kluwer Academic/ Plenum Publishers. 2001.

George C. Denniston, Frederick Mansfield Hodges, and Marilyn Fayre Milos, eds. *Flesh and Blood: Perspectives on the Problem of Circumcision in Contemporary Society.* New York: Kluwer Academic/Plenum Publishers. 2004.

Efua Dorkenoo. *Cutting The Rose: Female Genital Mutilation: The Practice and its Prevention.* London: Paul & Co Pub Consortium. 1996.

Paul M. Fleiss and Frederick M. Hodges. *What Your Doctor May Not Tell You About Circumcision.* New York: Warner Books. 2001.

Leonard B. Glick. *Marked in Your Flesh: Circumcision from Ancient Judea to Modern America.* New York. Oxford University Press. 2005.

Ronald Goldman, PhD. *Circumcision: The Hidden Trauma.* Boston: Vanguard. 1996.

Ronald Goldman, PhD. *Questioning Circumcision: A Jewish Perspective.* Boston: Vanguard. 1997.

David L. Gollaher, PhD. *Circumcision: A History of the World's Most Controversial Surgery.* New York: Basic Books. 2000.

Pia Grassivaro Gallo and Franco Viviani, eds. *Female Genital Mutilation: A Public Health Issue Also in Italy.* Proceedings of the 1994 International Symposium on Female Genital Mutilation, May 3rd, 1994, Padua, Italy. Padua: UNIPRESS. 1995.

Gruenbaum, Ellen. The Female Circumcision Controversy. Philadelphia: University of Pennsylvania Press, 2000.

Fran P. Hosken. *The Hosken Report: Genital and Sexual Mutilation of Females.* 4th edition. Lexington, MA: Women's International Network News. 1994.

Hanny Lightfoot-Klein. *Prisoners of Ritual: An Odyssey Into Female Genital Circumcision in Africa.* New York: Harrington Park Press. 1989.

Hanny Lightfoot-Klein. *Secret Wounds.* Bloomington: 1stBooks, 2002. [ISBN 1-4033-9699-X].

Peggy O'Mara, ed. *Circumcision: The Rest of the Story.* Santa Fe, NM: Mothering.1993.

Thomas J. Ritter, MD, and George C. Denniston, MD. *Doctors Re-examine Circumcision.* 3rd Edition. Aptos, CA: Hourglass. 2002.

Margaret Somerville. *The Ethical Canary: Science, Society and the Human Spirit.* Toronto: Penguin Books. 2000.

Edward Wallerstein. *Circumcision: An American Health Fallacy.* New York. Springer. 1980.

Kayla Weiner, PhD, and Arinna Moon, MA., eds. *Jewish Women Speak Out: Expanding the Boundaries of Psychology.* Seattle, WA: Canopy Press. 1995.

4. VIDEOTAPES/FILMS

Circumcision? Intact Facts. 18-min. VHS. $44.05. Injoy Productions, 1435 Yarmouth, Suite 102-B, Boulder, CO 80304. Tel: 800-326-2082.

The 8th Day. 53 min. VHS. $30ppd in US, $50ppd outside US. Keren Markuze, POB 361425, Los Angeles, CA 90036. Tel: 323-936-6802.

Facing Circumcision: Eight Physicians Tell Their Stories and Reveal the Ethical Dilemmas of Physicians who Circumcise Newborns. Nurses for the Rights of the Child. 20 minutes. VHS. 1998. Nurses for the Rights of the Child. 369 Montezume #354, Santa Fe, New Mexico, 87501. 505-989-7377. www. cirp.org/nrc/.

Fire Eyes. Soraya Mire. 60 minutes. sale: 16mm $2,000, video $445, rental: 16mm $300, video $85. Filmakers Library, Inc., 124 East 40th Street, New York, Y 10016 (212-808-4980).

It's a Boy! Victor Schonfeld. 41 minutes. VHS. $295 institutions, $195 individuals, $65 rental. Filmmakers Library, 124 East 40th Street, New York, NY 10016. el: 212:808-4980. Fax: 212-808-4983.

The Nurses of St. Vincent: Saying "No" To Circumcision. Barry Ellsworth. Includes NOHARMM Rally at California Medical Association. VHS. 30 minutes. 29. NOCIRC, POB 2512, San Anselmo, CA 94979.

Tahara. Sara Rashad. 18-min. NTSC VHS. $30 ($200 for institutions, schools, libraries). Order at www.tahara-film.com.

Whose Body, Whose Rights? Lawrence Dillon and Tim Hammond. 1996. 56 minutes. VHS. Home Sales: Video-Finders, 1-800-343-4727. Educational Sales: $195, Rental $70, Catalogue no. 38342, University of California Extension, Center for Media and Independent Learning, 2000 Center Street, Fourth Floor, Berkeley, CA 94704, 510-642-0460.

5. NEWSLETTERS

NOCIRC Annual Newsletter. Marilyn Fayre Milos, RN, Editor. NOCIRC. POB 2512. San Anselmo, CA 94979-2512. USA.

NOCIRC of Michigan Informant, Norm Cohen, Editor. POB 333, Birmingham, MI 48012.

NORM NEWS, David Smith, Editor, Send SASE and Postal Reply Coupon. NORM-UK, POB 71, Stone, Staffordshire ST15 0SF, England.

INDEX